Fresh Approaches to Brick Production and Use in the Middle Ages

Proceedings of the session
'Utilization of Brick in the Medieval Period –
Production, Construction, Destruction'

Held at the European Association of Archaeologists (EAA)
Meeting, 29 August to 1 September 2012 in Helsinki, Finland

Edited by

Tanja Ratilainen
Rivo Bernotas
Christofer Herrmann

BAR International Series 2611
2014

Published in 2016 by
BAR Publishing, Oxford

BAR International Series 2611

Fresh Approaches to Brick Production and Use in the Middle Ages

ISBN 978 1 4073 1242 2

© The editors and contributors severally and the Publisher 2014

The authors' moral rights under the 1988 UK Copyright,
Designs and Patents Act are hereby expressly asserted.

All rights reserved. No part of this work may be copied, reproduced, stored,
sold, distributed, scanned, saved in any form of digital format or transmitted
in any form digitally, without the written permission of the Publisher.

BAR Publishing is the trading name of British Archaeological Reports (Oxford) Ltd.
British Archaeological Reports was first incorporated in 1974 to publish the BAR
Series, International and British. In 1992 Hadrian Books Ltd became part of the BAR
group. This volume was originally published by Archaeopress in conjunction with
British Archaeological Reports (Oxford) Ltd / Hadrian Books Ltd, the Series principal
publisher, in 2014. This present volume is published by BAR Publishing, 2016.

Printed in England

BAR titles are available from:

	BAR Publishing
	122 Banbury Rd, Oxford, OX2 7BP, UK
EMAIL	info@barpublishing.com
PHONE	+44 (0)1865 310431
FAX	+44 (0)1865 316916
	www.barpublishing.com

TABLE OF CONTENT

List of Contributors and Editors in Alphabetical Order	ii
Preface	v
Early Medieval Brickmaking: a Cross-Channel Perspective Based on Recent Luminescence and Archaeomagnetic Dating Results Sophie Blain, Philippe Lanos, Ian Bailiff, Pierre Guibert, Christian Sapin	1
Brick Production and Brick Building in Medieval Flanders Vincent Debonne	11
Approach on Brick and its Use in Brussels from the 14th to the 18th Century Philippe Sosnowska	27
The Use of Early Bricks in Secular Architecture of the 12th and Early 13th Century in Lübeck Dirk Rieger	39
The Origin and Rise of Brick Technology and Use in Medieval Pomerania, Pomerelia and Lower Silesia Felix Biermann and Christofer Herrmann	51
Brick in Historic Architecture: The Research Experience of Warsaw Włodzimierz Pela	61
Facts and Thoughts on Medieval Brick Building in the Northern Periphery of Europe Gunilla Malm	71
A Recently Excavated Medieval Brick Kiln at Saltsjö Boo, Sweden and the HXRF Analyses of its Products Jan Peder Lamm and Anders Lindahl	81
Unfired Bricks Used for a Medieval Oven in Turku, Finland Tanja Ratilainen	93

List of Contributors and Editors in Alphabetical Order

Bernotas, Rivo, BA

Is an excavation manager working for the Lahti City Museum in Finland. He is also a doctoral student in the Department of Archaeology, University of Turku, Finland. His research interests lie in medieval urban archaeology and social history, covering topics like town fortifications, infrastructure, buildings, waste management and building materials. He has also worked extensively on applying the dendrochronological dating method to urban archaeology. He is currently working on his doctoral thesis on the town fortifications in the Northern Baltic Sea region, concentrating on the towns of medieval Livonia. Contact: Department of Archaeology, Henrikinkatu 2, 20014, University of Turku, Finland, e-mail: rivobernotas@gmail.com

Biermann, Felix, Docent

is a Heisenberg-Stipendiat of German Research Foundation at Georg-August-University Göttingen, prehistoric branch. His research interests are the medieval archaeology of middle and eastern Europe, and the history and archaeology of castles, towns, villages and monasteries. Contact: Georg-August-University Göttingen, Seminary of Prehistoric Archaeology, Nikolausberger Weg 15, D-27073 Göttingen, Germany, e-mail: Felix.Biermann@phil.uni-goettingen.de

Blain, Sophie, Dr

is currently a postdoctoral researcher F.R.S.-FNRS at the University of Liège, Belgium. Sophie has a dual PhD in Archaeological Sciences from the Universities of Durham (UK) and Bordeaux (France). She has been working on a number of standing medieval churches in south-eastern England, north-western France, Belgium and northern Italy and, from now on, in Burgundy, France. The originality of her approach to these monumental objects of study lies in its transdisciplinary methodology which combines building archaeology and scientific dating methods applied to building materials, such as dendrochronology on wood beams and luminescence (TL/OSL) dating methods on ceramic building materials and mortar. Her work is funded by Fonds de la Recherche Scientfique - National Fund for Scientific Research (F.R.S.-FNRS). Contact: Bât B5A, Centre Européen d'Archéométrie Université de Liège, Allée du 6 Août, 17, 4000 Liège, Belgium, e-mail: blain.sophie@gmail.com

Debonne, Vincent, MA

is a Built Heritage researcher at the Flanders Heritage Agency (Belgium), and has a Masters in Art History and in Conservation of Historic Towns and Buildings. He is currently a doctoral student at the Katholieke Universiteit Leuven (Belgium). His research interests are the architecture of medieval Flanders and building archaeology. He is the holder of a Special Ph.D. Fellowship, granted by the Fonds Wetenschappelijk Onderzoek – Vlaanderen. Contact: Koning Albert II-laan 19 bus 5, 1210 Brussel, Belgium, e-mail: vincent.debonne@rwo.vlaanderen.be

Herrmann, Christofer, Professor

is a professor at the Institute of Art History at Gdansk University, Poland. His research interests are medieval architecture and brick architecture of the Baltic Sea, and financing and building organization in Middle Ages. Contact: e-mail: chriherr@yahoo.de

Lamm, Jan Peder, Docent

is retired Director of Research at the Historical Museum in Stockholm and specialized in the archaeology of the Migration period. The Scandinavian gold jewelry of this period has been in focus of his interest. At the moment, together with German colleagues, he works on a monograph of the famous three gold collars found in Sweden, which hopefully will be published in 2014. Destiny in form of a well located ruin of a brick kiln has enabled him after retirement also to enter the realms of medieval archaeology. Contact: e-mail: janpeder.lamm@bredband.net

Lindahl, Anders, Professor

is a professor in Laboratory and Experimental Archaeology at Lund University and head of the Laboratory for Ceramic Research, Department of Geology. He has worked with ceramic analyses for more than a quarter of a century on materials from a broad range of periods and regions. His work is mainly on basic research within documentation and methods to analyse ceramic materials found at archaeological excavations. He has special research foci on African pottery, ethnographic studies and experimental pottery making. Contact: Laboratory for Ceramic Research, Department of Geology, Sölvegatan 12, 223 62 Lund, Sweden, e-mail: Anders.Lindahl@geol.lu.se, home page: http://www.geol.lu.se/kfl

Malm, Gunilla, BA

is a building archaeologist (senior), educated at Lund University. She worked as a First Antiquarian at the Swedish National Heritage Board (Raä) for almost 30 years and later at the Royal Swedish Academy of Letters, History and Antiquities. In the 1970s she took part in the archaeological excavation at Birka – one of the oldest towns or trade places of Sweden. Her main archaeological and building archaeological research is about Uppsala Cathedral and the archbishopric remains of Uppsala medieval Church Town as well as the prehistoric hillfort Gråborg of Öland – one of the largest prehistoric hillforts in Sweden. Contact: Kungsklippan 11, S 112 25 Stockholm, Sweden.

Pela, Włodzimierz, MA

is an independent researcher focusing on the archaeological and architectural study of medieval, post-medieval and modern towns (cities) in Poland. Contact: 01-703 Warszawa, Gąbińska 9/103, Poland, e-mail: wowkap@op.pl

Rieger, Dirk, Dr

is a field manager working for the Archaeological Service of the Hansestadt Lübeck. His research interest is mainly urban archaeology with a special focus on research on the genesis processes and their complex interactions with politics and economy in former Saxony. Contact: Projekt Großgrabung Gründungsviertel c/o 4.491 - Archäologie und Denkmalpflege, Abteilung Archäologie Hansestadt Lübeck, Braunstraße 21, 23552 Lübeck, Germany, e-mail: dirk.rieger@luebeck.de, http://www.AusgrabungGruendungsviertel.luebeck.de

Ratilainen, Tanja, LicPhil

is a PhD student in the Department of Archaeology, University of Turku, Finland. She is a building archaeologist specializing in medieval masonry buildings and interested in the stratigraphy of buildings, the history of bricks and tiles, and applying 3D techniques in buildings archaeology. She is currently working on her dissertation on the building process of the medieval Holy Cross (brick) church of Hattula in Häme, Finland. Her work is funded by the Kone Foundation. Contact: c/o Department of Archaeology, Henrikinkatu 2, 20014 University of Turku, Finland, e-mail: tanrat@utu.fi

Sosnowska, Philippe, Dr

is a doctor of History, Art and Archaeology, and currently a postdoctoral researcher. His main research topics are construction materials, and the study of both urban and rural dwellings in the Southern Netherlands, more specifically Brussels, from the 13th until the 19th century. His work is based on a double approach combining archaeology with construction history. His investigation fields cover on the one hand the issues of supply, manufacturing and technology, and on the other hand the evolution of habitats and the "ways" of living, by the analysis of former regulations, implementations and equipment, all comfort-related, and by the typological approach of dwellings. Contact: Centre de Recherches en Archéologie et Patrimoine, Université libre de Bruxelles • CP 175, Avenue F.D. Roosevelt, 50 B-1050 Bruxelles, Belgium, e-mail: philippe.sosnowska@ulb.ac.be

PREFACE

"Data! data! data! I can't make bricks without clay." These apt words were cried by Sherlock Holmes in "The Adventure of the Copper Beeches" by Sir Arthur Conan Doyle as he needed more information to draw conclusions. We hope that this publication in its way answers some of the questions which have troubled readers dealing with medieval bricks and, perhaps, provide fresh insights as well as raise new ideas and questions in the minds of discerning scholars such as Sherlock Holmes.

The core of this publication is formed of the papers presented in the session called "Utilization of Brick in the Medieval Period – Production, Construction, Destruction" held at the annual meeting of the European Association of Archaeologists in Helsinki in late summer 2012. With the kind approval of the British Archaeological Reports editors, we were also able to invite more contributors to complete this book.

We would like to thank all the contributors for their hard work and patience. We would like to express our gratitude to Turku Centre for Medieval and Early Modern Studies (TUCEMEMS) for the financial support, granted for formatting expenses of this publication. It saved a great deal of our time and effort (for finishing some of our PhDs) – thank you for the support. We would also like to thank the Finnish Doctoral Program in Archaeology for making it financially possible to organize the session in the EAA meeting.

We hope you enjoy reading the book.

The editors

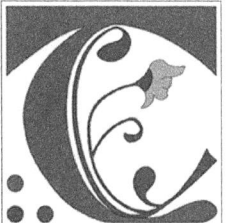

Turku Centre for
Medieval and Early
Modern Studies
tucemems.utu.fi

EARLY MEDIEVAL BRICKMAKING:
A CROSS-CHANNEL PERSPECTIVE BASED ON RECENT LUMINESCENCE AND ARCHAEOMAGNETIC DATING RESULTS

Sophie Blain[1], Philippe Lanos[2], Ian Bailiff[3], Pierre Guibert[4], Christian Sapin[5]

Abstract: The integration of dating methods in building archaeology has resulted in an advance in the qualitative and quantitative information available for the study of the history of architecture and building techniques. To examine the question of the origin of post-Roman ceramic building materials, archaeological studies of early medieval buildings in France and England have been combined with scientific dating methods and applied to their component tiles and bricks. The latter were sampled and analysed using luminescence dating techniques. The results show that, as well as the practice of reusing bricks or tiles salvaged from abandoned Roman sites, brickmaking was not a forgotten skill in north-western France and appears to have been continuously practised in the region. At the same time in England, the use of ceramic spolia was the rule and the reintroduction of post-Roman bricks in the island appears to be relatively late, although recent luminescence results indicate that this was earlier than the arrival of Cistercian builders in the late 12th century. Finally, the combined application of archaeomagnetism and luminescence to the dating of bricks has revealed unexpected and systematic evidence of an early medieval brickmaking process in north-western France.

Keywords: Anglo-Saxon, archaeomagnetism, building archaeology, Carolingian architecture, ceramic building materials, luminescence dating

Introduction

This work partly results from the studies carried out within the framework of a European Group of Research entitled 'Ceramic Building Material and dating methods' (2005–2012). It aimed at investigating further the history of early medieval architecture and building techniques. As such, it brought together a series of different experts to form a multidisciplinary team, including a building archaeological team, a history of art contribution and the laboratories of 'absolute' dating methods including radiocarbon dating, dendrochronology, archaeomagnetism and luminescence techniques. Ten laboratories were involved, located in United-Kingdom (University of Durham), France (Universities of Bordeaux, Dijon, Lyon, Paris, Rennes, Toulouse), Italy (Universities of Catania and Milan) and Belgium (University of Liège). This led to an innovative study in Europe that combined and correlated archaeological and archaeometric data.

This paper more particularly deals with the reassessment of the archaeological question related to the origin of ceramic building materials (CBM) and more specifically of bricks (this latter term is used to describe any CBM which walling with a minimum thickness of 2.5cm) in Anglo-Saxon and Carolingian churches (from the 7th to the 11th century AD): were they reused Roman material or contemporary to the early medieval construction? This question is worth rising since it is commonly known that brickmaking, introduced by the Romans to Britain and Gaul, was thought to have ceased following their departure and the subsequent arrival of Germanic settlers in the 5th century[6], to reappear only at the end of the 12th century mainly through the Cistercians[7]. In spite of this supposed multi-secular break in brickmaking, early medieval builders continued to use bricks, considered for a long time to be Roman reused material[8].

This paper provides an overview on the role of scientific dating methods of bricks and their contribution to our knowledge of early medieval brickmaking.

Aims

The use of ceramic building materials in a masonry structure is one of the main features of early medieval architecture, notably during the Carolingian and Anglo-Saxon periods, and this practice is visible in areas touched by Roman traditions, such as in north-western France and south-eastern England. These regions have often influenced each other due to their geographical proximity, which make them particularly receptive to foreign ideas and influences. Through trade, political and economic links, marriage (such as between the king of Kent Ethelbert (AD 560?–616) and Bertha, daughter of the Frank king Charibert) and activities

---

[1] Centre Européen d'Archéométrie - Université de Liège, Belgium, blain.sophie@gmail.com.
[2] IRAMAT-CRP2A – UMR 5060, CNRS – Université de Bordeaux 3, Géoscience-Rennes-Université de Rennes 1, France, philippe.lanos@univ-rennes1.fr
[3] Luminescence Laboratory, Department of Archaeology, University of Durham, South Road, DH1 3LE, UK ian.bailiff@durham.ac.uk.
[4] IRAMAT- CRP2A – UMR 5060, CNRS – Université de Bordeaux 3, Pessac, France, guibert@u-bordeaux3.fr.
[5] Laboratoire Artehis – UMR 5594, CNRS -Université de Bourgogne, Dijon, France, sapin.christian@wanadoo.fr.

[6] Lynch 1994, 3.
[7] Coomans and van Royen 2007, 1; Moore 1991.
[8] Morant 1768, 298; de Bouärd 1975, 55–6.

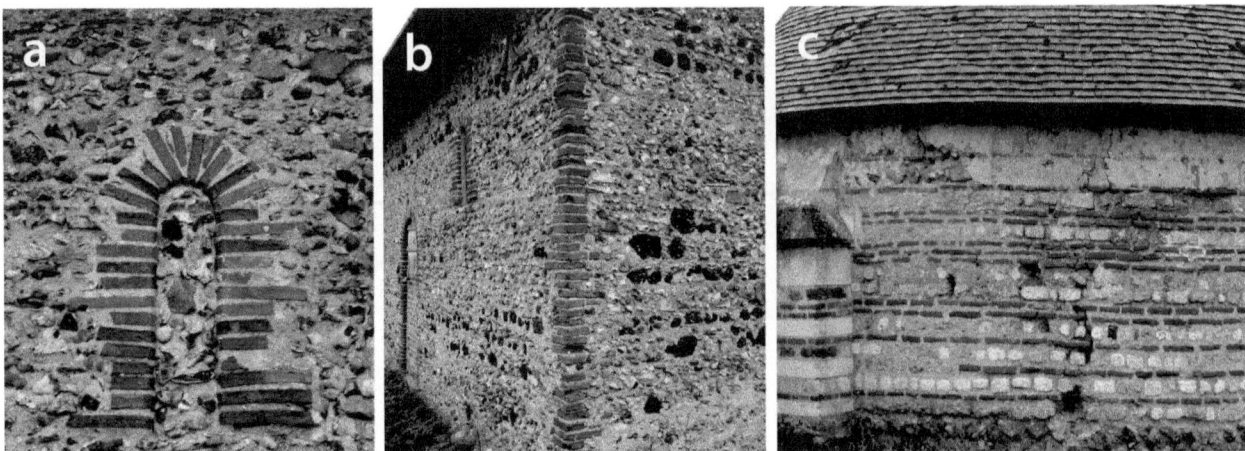

Figure 1. Bricks used in a window dressing (a) and a quoin (b) of Bradwell-juxt-Coggeshall church and in a row (c) within the wall of the apse of Rugles church (source: Sophie Blain).

of English scholars in the Carolingian Empire (e.g., Alcuin of York, counsellor of Charlemagne), technical knowledge and craftsmen were introduced in each region[9]. The case of the founder of the monasteries Monkwearmouth (674) and Jarrow (681), the Abbot Benedict Biscop (628–689), who called on Gallic craftsmen to build them more romano, reflects these close relationships.

Methodology

The study concentrates on standing buildings, showing large quantities of CBM, used homogeneously rather than sporadically and in an organised manner, as in voussoirs, in jambs of opening dressings (Figure 1a), in quoins (Figure 1b) and in rows (Figure 1c).

In France, this includes high status Carolingian churches such as the abbey church of St Philbert-de-Grandlieu near Nantes, the Collegiate church of St Martin in Angers, both in Pays-de-Loire region, the earliest surviving church of Notre-Dame-sous-Terre in the Mont-Saint-Michel (which constitutes the only standing element of the monastic origins of the abbey) and the more modest parish churches such as Vieux-Pont-en-Auge, Condé-sur-Risle or Rugles in Normandy, stylistically assigned to c. AD 1000[10]. In England, the selected buildings were the supposed oldest church of St Martin of Canterbury in England which is still in use[11], the late Anglo-Saxon parish churches of Lower Halstow[12] and Darenth[13] in Kent, and in Essex the only surviving Anglo-Saxon architectural remains of Colchester: the tower of Holy Trinity[14] and for the later building, the Saxo-Norman church of St Martin-of-Tours in Chipping Ongar, near Chelmsford. Figure 2 shows the location of these sites.

Figure 2. Location of the sites included in the study (source: Sophie Blain).

In order to investigate early medieval brickmaking, the bricks of these buildings were analysed with two scientific dating methods: thermoluminescence / optically stimulated luminescence (TL/OSL) and archaeomagnetism (AM). For each of these methods, the dating process starts in the field. The bricks were sampled either with a chisel and a hammer or with a diamond-faced core drill designed for wet-cutting. Aesthetic damage was minimized by repairing the hole with pigmented mortar as close to the original brick colour as possible. The samples were then mechanically and chemically prepared and analysed in the laboratories

---

[9] Levison 1946; Webster and Blackhouse 1991.
[10] Baylé 1992, 43; 1999; 2000, 7–8.
[11] Colgrave and Mynors 1969; Tatton-Brown 1985, 14; Bell 2005, 124.
[12] Baldwin-Brown 1925, 426, 456; Olive 1918, 157.
[13] Elliston-Erwood 1912, 83.
[14] Laing 1848, 20.

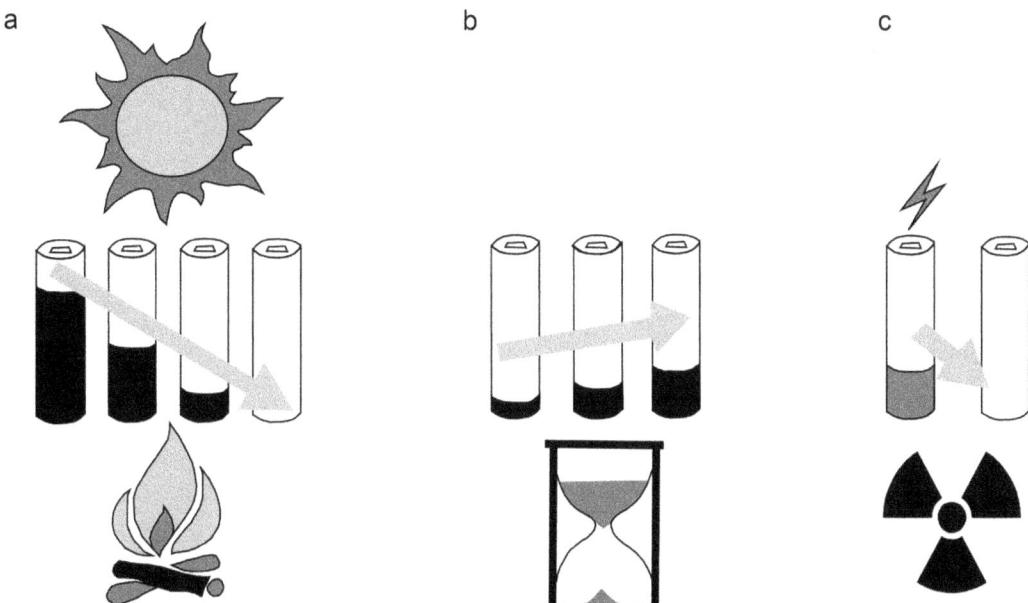

Figures 3. The crystals within the material to be dated can be compared to (dis)charging batteries: a) first, the heat or light exposure reset the material which causes the discharging of the battery; b) then, with the effects of the surrounding radioactivity lead to the continuous filling of the crystal defects, as such as a battery recharge; c) finally, in the laboratory, when applying an external stimulation to the sample, the defects are emptied[17] (source: Sophie Blain).

according to the requirements of the techniques[15]. Both methods date the same event, which is the last firing of the bricks, but they are based on different physics principles.

TL/OSL

Phenomenon

The luminescence method uses the property of irradiated crystals, such as quartz, that emit light (luminescence) when stimulated by heat or light[16]. This light emission results from the cumulative effects of natural irradiation on crystals that are typically found in the ceramic materials (the radiation includes α, β particles and γ rays emitted by naturally occurring radionuclides and cosmic rays). During irradiation, electrons are liberated from their parent atom in the crystal and are trapped in the crystal defects with the number of trapped electrons increasing with time. The application of heat or light liberates the trapped electrons, leading to the emission of light, the intensity of which is proportional to the number of trapped electrons and ultimately to the amount of radiation received and, assuming the continuity and constancy of the irradiation, it is also proportional to the duration of the exposure (i.e. the age). The mechanism underlying the method can be illustrated using an analogy based on the charging and discharging of batteries (Figure 3).

Principle of the method

In the case of bricks, the last archaeological firing of the ceramic empties the traps of the crystal filled previously (known as a zeroing event); in the metaphoric image of the battery, this correspond to its discharging (Figure 3a). The process of electron trapping recommences after firing allowing the subsequent measurement of luminescence which results only from after the zeroing event. The battery is then charging (Figure 3b). The quantity of absorbed and stored energy is called paleodose and is proportional to the time elapsed since firing and the radioactive dose rate. During laboratory experimentation traps are emptied by the application of light or high temperature which, as previously stated, results in the further emission of light (of different wavelength). This process is called Optically stimulated luminescence (OSL) or Thermoluminescence (TL), which could be compared with the trigger mechanism of the battery discharging (Figure 3c).

The luminescence age was calculated using the age equation below:

Age (years) = Paleodose (mGy) / Annual dose (mGy/yr).

Archaeomagnetism

The principles of archaeomagnetism (AM) are based on the fact that the Earth behaves like a magnet producing magnetic fields called Earth Magnetic Fields (EMF) (Figure 4a) and on the property of ceramic material to record this field.

---

[15] Blain 2011, 28–45.
[16] Aitken 1985.
[17] Duller 2008.

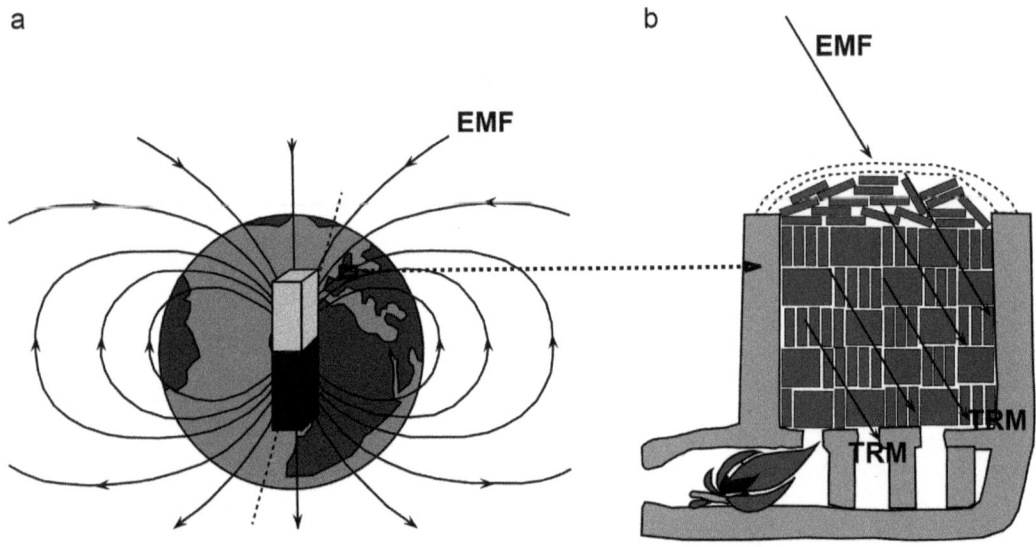

Figures 4. a) Schematic representation of the magnetic fields around the planet; b) Recording of the EMF during the cooling at the end of the brick firing (source: Sophie Blain & Philippe Lanos).

Phenomenon

The direction (inclination, declination) and intensity of the Earth Magnetic Field (EMF) vector evolves through time. At the end of the firing process of brickmaking, i.e. during the cooling, the ceramics reach down the 'temperature of Curie'. The iron oxides within the clay acquire then a remanent magnetization which direction is parallel to the direction of the ambient EMF at the place of this cooling and which intensity is proportional to the EMF one (Figure 4b). This very stable magnetization known as 'thermoremanent magnetization' (TRM; also a vector) is kept unchanged through time if the material is not fired again, and therefore provides a record of the past geomagnetic field.

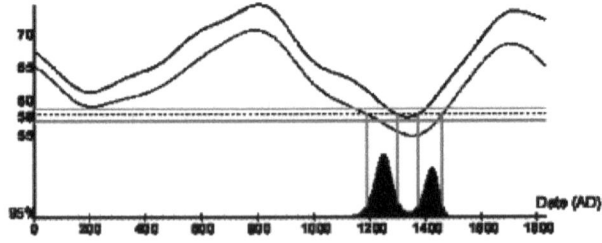

Figure 5. AM Inclination dating: comparison of the brick thermoremanent magnetization and the reference curve (source: Lanos & Philippe Dufresne).

Principle of the method

The measurement of this magnetization on a group of bricks (sampled considering their orientation in space[18]) fired at a same time in a kiln allows dating the firing of the making, by comparing the TRM direction and intensity with reference secular variation curves of the EMF direction, known for the area where the sample was found. The date is obtained by the combination with the measured archaeomagnetic direction of the sample and a dated segment of the reference curve of the geomagnetic field (Figure 5).

Results

The results derived from the application of these two methods on bricks sampled from the sites of our study have three main aspects: the origin of the bricks used in early medieval buildings, the reintroduction of the know-how in post-Roman England and the early medieval brickmaking process in France.

Roman spolia versus newly made products

The results of luminescence dating applied to bricks show that brickmaking was not a forgotten skill in France[19]. On the contrary, it seems to have been continuously practised in north-western France at least from the 7th century – since no earlier building with bricks was available or suitable for this analysis (Figure 6).

This is in agreement with the earliest evidence of brickmaking for Normandy provided by the early medieval French historian, Dudo of St Quentin (AD 960–1026), who, concerning the rebuilding of the sanctuary of la Trinité at Fécamp by the Duke Richard 1st (consecrated in AD 990), mentioned that CBM were made for the building: 'lateribus artificialiter cumpotis'[20]. However we have no indication whether these materials were wall bricks, roof or floor tiles. Furthermore, recent studies on bricks from historic monuments in the Lyons and Bordeaux regions have also

---

[18] Lanos et al. 2008, 103.

[19] Blain et al. 2007; 2009; Blain 2011; Guibert et al. 2009
[20] Lair 1865, 290–291; Renoux 1991, 473–5.

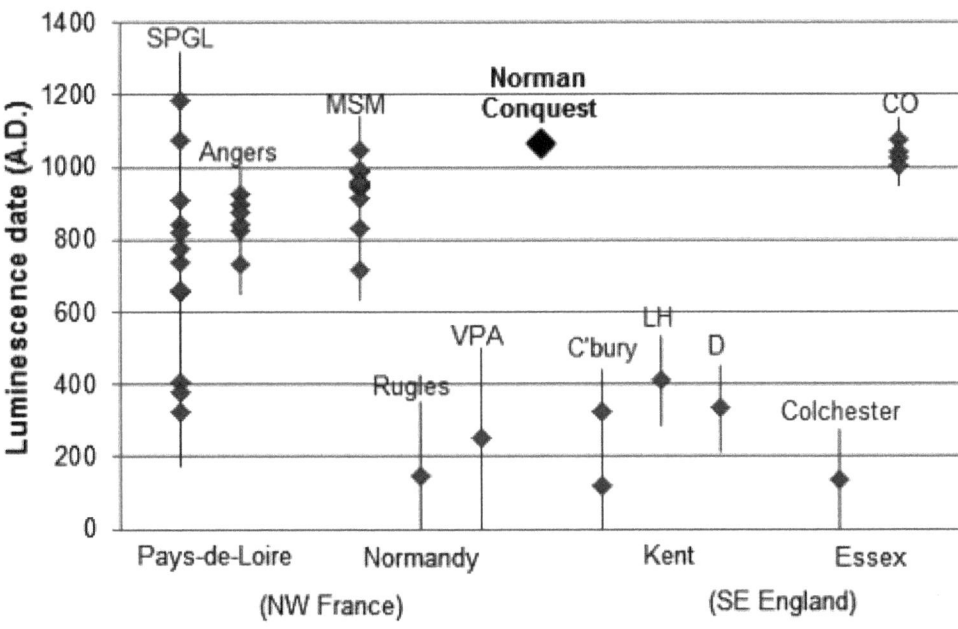

Figure 6. Individual luminescence dates geographically grouped. The vertical bars correspond to the overall error (±1σ). SPGL refers to Saint-Philbert-de-Grandlieu; MSM to Mont-Saint-Michel; VPA to Vieux-Pont-en-Auge; C'bury to Canterbury; LH to Lower Halstow; D to Darenth; CO to Chipping Ongar (source: Sophie Blain).

provided evidence of a continuity of brick production during the early Middle Ages[21].

However it should be noted that new materials were produced mainly for high status churches, such as in St Philbert-de-Grandlieu, St Martin's in Angers, Notre-Dame-sous-Terre's in Mont-Saint-Michel, since it would have been a time consuming and expensive process that only the commissioners of these buildings could have afforded. Reusing bricks or tiles from salvaged abandoned sites was also practiced in France but mainly for more modest buildings, i.e. the parish churches such as Rugles, Vieux-Pont-en-Auge (Figure 6), and likely to have been for economic reasons. Bouvier et al.[22] have also shown the practice of using both new and reused materials in a same building phase for high status buildings in the Lyons area.

In France it appears that both practices, the use of spolia and contemporary brickmaking coexisted during this period, yet at the same time, in England, the use of ceramic spolia appears to have been invariably practised as confirmed by luminescence dating results[23] (Figure 6) and cautious field survey[24]. This is often noticeable by the use of roof tiles diverted from their primarily function, for instance imbrices and tegulae incorporated within walls, or blocks of opus signinum (sometimes even still attached to a reused brick) used as rubble.

Given this long tradition of material reuse in south eastern England, the question of when the reintroduction of brickmaking occurred is worth raising.

The reintroduction of post-Roman brickmaking in England

Hitherto, the occasional local production of great bricks from Coggeshall abbey at the end of the 12th century has been considered as the first post-Roman production of bricks in England[25]. In AD 1140, Mathilda of Boulogne (AD 1105?–52), Queen consort of England (AD 1135–52), sold to the Savignac Order the field where the abbey was to be founded in AD 1167[26]. She has also been related by marriage to the manor of Chipping Ongar since the latter was offered in AD 1125 as a dowry for her wedding with Stephen, Count of Blois and future King of England[27]. There are currently no known surviving historic sources related to the origins of the Chipping Ongar church building, but stylistically it is attributed to the early Norman period[28]. Despite having been described as Roman reused material in the architectural records of the church[29], likely to be due to the broken state of a number of them, the bricks were selected for luminescence dating tests given their non-Roman-like features: e.g. sandy texture, larger size, similar to early medieval material French bricks such as those of St-Philbert-de-Grandlieu, Angers, Mont-Saint-Michel. Although, due to the connection between the two ecclesiastical buildings, through Mathilda, a provenance from Coggeshall for the CBM of Chipping Ongar was considered plausible, the bricks of Chipping Ongar are not of comparable size to the Coggeshall 'great bricks', being

---

[21] Bouvier et al. 2013; in press.
[22] Bouvier et al. 2013.
[23] Bailiff et al. 2010; Blain 2011.
[24] Ryan 1996; Blain 2011.
[25] Gardener 1955; Rodwell 1998; Coomans and van Royen 2007.
[26] Gardener 1955.
[27] Morant 1768; Powell 1956, 159.
[28] RCHM 1921, 51; Powell 1956, 162–3; Pevsner and Radcliffe 1965, 125; SMR 4110
[29] RCHME 1921, 51; Powell 1956, 163; Pevsner and Radcliffe 1965, 124.

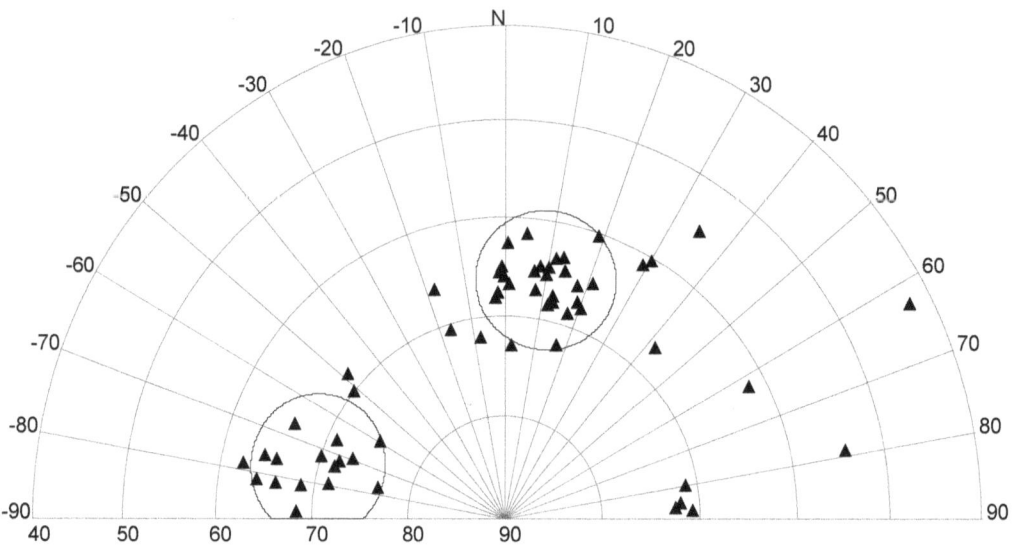

Figure 7. Stereogram of magnetization (source: Philippe Lanos & Philippe Dufresne).

longer and larger but slightly thinner. The dimensions of the Coggeshall bricks are 320–30 x 150–60 x 45–50mm and those of the Chipping Ongar bricks are 380 x 190 x 38mm[30].

OSL dates were obtained for four bricks from Chipping Ongar church and using this mean date value of AD 1040±30 calculated for their manufacture[31] (Figure 6). This result suggests the existence of local brickmaking approximately contemporary with the Norman Conquest and is particularly interesting as previously no wall brick of this date had been identified in England. Furthermore, Inductively Coupled Plasma – Atomic Emission Spectrometry (ICP-AES) analyses performed on local clay in Essex and on bricks from nearby sites included Chipping Ongar suggest the latter may have been made at the site of the church itself, or nearby[32].

This provides a strong indication that the bricks tested are so far the earliest known post-Roman bricks produced in England. Several scholars have already expressed reservations concerning the pioneering role of Cistercians in the technological reintroduction of brickmaking in England[33] and this new dating evidence supports such doubts.

Early medieval brickmaking process in north-western France

Archaeomagnetism can provide evidence of the positioning of the bricks during their firing by reconstructing the relative orientation of the inclination vectors[34]. Indeed it is possible to express the firing position of the bricks with respect to their common magnetic north[35]: the angle

Figure 8. Grid pattern of the bricks' position (source: Philippe Lanos & Philippe Dufresne).

between one vertical plane of the brick (i.e. flat plane in the case of the upright position) and the magnetic north at the time of firing is named 'deviation'.

For instance for the Roman period, the stereogram of these values (inclination, deviation) displays two orthonormal groups (Figure 7), indicating a filling of the kiln with a grid pattern[36] (Figure 8).

However in the case of early medieval bricks dated by TL/OSL, the stereogram of the directions of magnetization indicates rather a more disorganised filling[37], such as with the medieval bricks of St Martin's in Angers[38] (Figure 9).

---

[30] Ryan 1996; Rodwell 1998.
[31] Bailiff et al. 2010; Blain 2011.
[32] Hughes 2009.
[33] Ryan 1996; Rodwell 1998; Andrews 2008.
[34] Lanos et al. 2008.
[35] Lanos et al. 2008, 104–5.

[36] Lanos 1990; 1994; 2002.
[37] Lanos et al. 2008, 103.
[38] Blain et al. 2011.

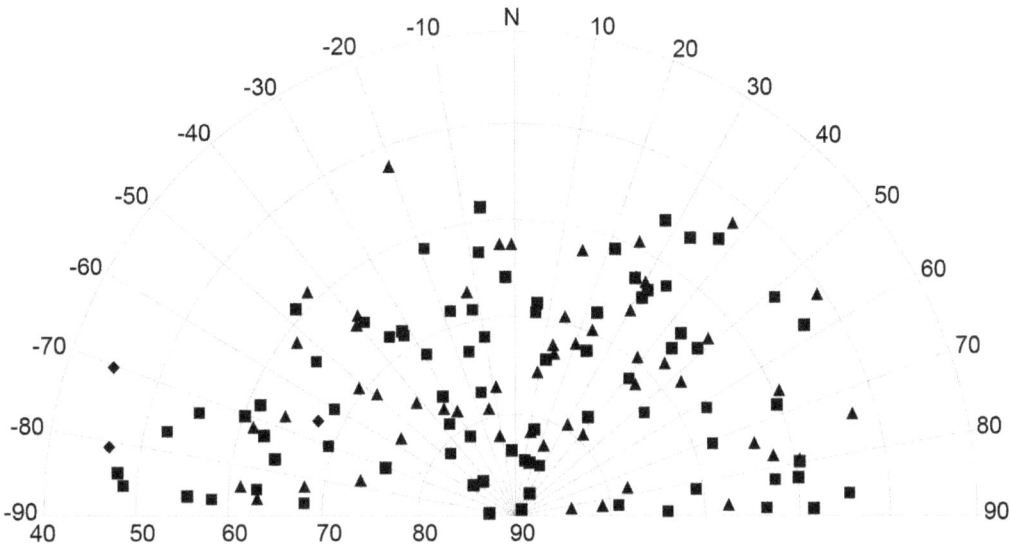

Figure 9. Stereogram of magnetization directions measured on a set of bricks from St Martin's Angers: results obtained after demagnetization and separation of the primary component corresponding to the production (firing) of the bricks (source: Philippe Lanos & Philippe Dufresne).

To explain this phenomenon, two assumptions have been raised. The first one suggests a firing in a clamp without kiln walls which would induce a random tilting of the bricks[39]. The second assumption suggests an internal settling of the filling of bricks within the chamber of the kiln during or at the end of the firing due to the shrinking of some bricks according to the degree of heating. Such an internal movement has an impact on the final archaeomagnetic vector directions and should give a large scattering of the magnetization directions in the stereogram (Figure 9).

Discussion

This study has allowed a revaluation of the commonly held view that bricks used in early medieval buildings were reused[40]. Furthermore it has shown that the combination of brick dating by luminescence and the analysis of their position during firing using archaeomagnetism can provide indirect evidence of the type of structure used in the early medieval brickmaking process. The hypothesis of a temporary kiln could explain why no brick kiln has been archaeologically detected for the early medieval period, giving rise to the established assumption that no bricks were produced in this time. Finally this study indicates that a reassessment of the role of Cistercians in the reintroduction of brickmaking in England is required and moreover prompts the same fundamental question regarding the reintroduction of post-Roman brickmaking technology for other areas of north-western Europe.

Acknowledgements

Thank you to Philippe Dufresne, engineer at the Universities of Rennes 1 and Bordeaux 3, and Christian Comte, Master student of the University of Bordeaux 3, for their technical contribution to the study.

This research was supported by the CNRS, Department of Humanities and Social Sciences within the frame of the GdRE (Groupe de Recherche Européen) 'Terres Cuites Architecturales et nouvelles méthodes de datation', the Universities of Bordeaux 3, Lyon and Rennes (France), the Conseil Général du Maine-et-Loire and by the Aquitaine Region Council (PhD grant), the bilateral programmes of Alliance between UK and France and Tournesol between France and Belgium.

Thank you to the Universities of Bordeaux 3 (F), Dijon (F), Rennes 1 (F), Lyon (F), Durham (UK), Milan (I), Catane (I), Liège (B) for their support during the execution of the projects.

References

Bailiff, I.K., Blain, S., Graves, C.P., Gurling, T., Semple, S. 2010. Uses and recycling of brick in medieval and Tudor English buildings: insights from the application of luminescence dating and new avenues for further research. The Archaeological Journal 167, 165–196.

Baldwin-Brown, G. 1925. The Arts in Early England, Anglo-Saxon Architecture. Murray, London.

Baylé, M. 1992. Les origines et les premiers développements de la sculpture romane en Normandie. Art de Basse-Normandie, n°100 bis.

Baylé, M. 1999. Relations techniques et formelles entre l'architecture religieuse et l'architecture civile, aux XIe et XIIe siècles. La Maison Médiévale en Normandie et en Angleterre, Proceedings of the tables rondes de Rouen et de Norwich (1998–1999). Société libre d'émulation, 31–39.

---

[39] Lanos et al. 2008, 105.
[40] Blain 2011.

Baylé, M. 2000. Norman Architecture towards the Year 1000. Proceedings of the Battle Conference (Anglo-Norman Studies) 1999, Woodbridge.

Blain, S., Guibert, P., Bouvier, A., Vieillevigne, E., Bechtel, F., Sapin, C., Baylé, M. 2007. TL-dating applied to building archaeology: The case of the medieval church Notre-Dame-Sous-Terre (Mont-Saint-Michel, France). Radiation Measurements 42, 1483–1491.

Blain, S., Bailiff, I.K., Guibert, P., Bouvier, A., Baylé, M. 2010. An intercomparison study of luminescence dating protocols and techniques applied to medieval brick samples from Normandy (France). Quaternary Geochronology 5, 311–316.

Blain, S., Guibert, P., Prigent, D., Lanos, P., Oberlin, C., Sapin, C., Bouvier, A., Dufresne, P. 2011. Dating methods combined to building archaeology: the contribution of thermoluminescence to the case of the bell tower of St Martin's church, Angers (France). Geochronometria 38, 55–63.

Blain, S. 2011. Les terres cuites architecturales des églises du haut Moyen Age du NW de la France et du SE de l'Angleterre. Application de la datation par luminescence à l'archéologie du bâti. BAR International Studies 2189, BAR Publishing, Oxford, 221p.

Bell, T. 2005. The Religious Reuse of Roman Structures in Early Medieval England. BAR British Series 390, BAR Publishing, Oxford.

Boüard, M. (de). 1975. Manuel d'archéologie médiévale: de la fouille à l'histoire. collection Regards sur l'Histoire, Paris.

Bouvier, A., Guibert, P., Sapin, C., Blain, S.. Apports de la luminescence à la chronologie de la basilique Saint-Seurin de Bordeaux. Association Française d'Archéologie Mérovingienne. Aquitania, in press.

Bouvier, A., Guibert, P., Blain, S., Raynaud, J.F. 2013. Interdisciplinary study of the Early Building Phases of St Irénée's Church (Lyon, France): The Contribution of Luminescence Dating. Archeosciences 37, 155–171.

Colgrave, B. and Mynors, R.A.B. (ed). 1969. Bedae Venerabilis Historia Ecclesiastica Gentis Anglorum. Oxford.

Coomans, Th. and van Royen, H. 2007. L'architecture médiévale en briques en Flandre et dans le nord de l'Europe: la question de l'origine cistercienne. Colloque international à l'abbaye des Dunes (24–26 octobre 2007), Cîteaux. Commentarii Cistercienses, Revue d'Histoire Cistercienne, t.58, fascicule 3–4, 299–304.

Duller, G.A.T. 2008. Luminescence Dating: guidelines on using luminescence dating in archaeology. Swindon, English Heritage.

Elliston-Erwood, F.C. 1912. The Architectural History of the Church of St Margaret, Darenth. Proceedings of the Woodwich Antiquarian Society 17, 83–94.

Gardner, J.S. 1955. Coggeshall Abbey and its Early Brickwork. Journal of the British Archaeological Association 18, 19–32.

Guibert, P., Bailiff, I.K., Blain, S., Gueli, A.M., Martini, M., Sibilia, E., Stella, G., Troja, S. 2009. Luminescence dating of architectural ceramics from an early medieval abbey: the St-Philbert intercomparison (Loire Atlantique, France). Radiation Measurements 44, 488–493.

Hughes, M.J. 2009. ICP-AES analysis of ceramic building material. In: Drury, P. and Simpson, R.W. Hill Hall: A Singular House Devised by a Tudor Intellectual. London, Society of Antiquaries of London, 366–379.

Lanos, Ph., Chauvin, A., Dufresne, Ph. 2008. Terre cuite architecturale et archéomagnétisme. in: Sapin, C., Baylé, M., Büttner, S., Guibert, P., Blain, S., Lanos, P., Chauvin, A., Dufresne, P., Oberlin, C. 2008. Archéologie du bâti et archéométrie au Mont-Saint-Michel, nouvelles approches de Notre-Dame-sous-Terre. Archéologie Médiévale 38, 102–108.

Laing, G.E. 1848. Saxon Tower of Trinity Church, Colchester. Journal of British Archaeological Association 3, 19–22.

Lair, J. (ed.). 1865. Dudon de Saint-Quentin, De moribus et actis primorum Normanniae ducum. M.S.A.N., t. XXIII, Caen, Le Blanc-Hardel.

Lanos, Ph. 1990. La datation archéomagnétique des matériaux de construction d'argile cuite, apports chronologiques et technologiques. Gallia 47, 321–341.

Lanos, Ph. 1994. Pratiques artisanales des briquetiers et archéomagnétisme des matériaux d'argile cuite. Une histoire de positions de cuisson. Histoire & Mesure IX-3/4, 287–304.

Lanos, Ph. 2002. Sites du Château Aurélien et du Gargalon: analyses archéomagnétiques de deux lots de briques, chap.7: chantiers et bâtisseurs. In: L'aqueduc romain de Fréjus, sa description, son histoire et son environnement, in: Gébara, C. and Michel, J.M. (dir). Revue Archéologie de Narbonnaise, supplément 31, 225–233.

Levison, W. 1946. England and the Continent in the Eight Century, Clarendon Press, Oxford.

Lynch, G. 1994. Brickwork. History, Technology and Practice. vol 1, Donhead, London.

Moore, N.J. 1991. Brick. in: Blair, J. and Ramsay, N. (ed). English Medieval Industries. Craftsmen, Techniques, Products. The Hambledon Press, London, 211–236.

Morant, P. 1768. The History and Antiquities of the County of Essex, II, T. Osborne; J. Whiston; S. Baker; L. Davis and C. Reymers; and B. White, London.

Olive, E.R. 1918. Lower Halstow church. Archeologia Cantiana 33, 157–166.

Pevsner, N. and Radcliffe, E. 1965. The Buildings of England. Essex, Penguin Books, London.

Powell, W.R. (ed.). 1956. A history of the county of Essex 4. Victoria history of the counties of England, London, Constable.

Royal Commission on the Historical Monuments of England (RCHME). 1921. An inventory of the historical monuments in Essex, Central and South-West vol. 2, HMSO, London.

Renoux, A. 1991. Fécamp, du palais ducal au palais de Dieu: bilan historique et archéologique des recherches menées sur le site du château des ducs de Normandi : IIe siècle A.C.–XVIIIe siècle P.C., Editions du CNRS, Paris.

Rodwell, W. 1998. Holy Trinity church, Bradwell-juxta-Coggeshall: a survey of the fabric and appraisal of the

Norman brickwork. Essex Archaeological History 29, 59–114.

Ryan, P. 1996. Brick in Essex from the Roman Conquest to the Reformation. P. Ryan, Chelmsford.

Sapin, C., Baylé, M., Büttner, S., Guibert, P., Blain, S., Lanos, P., Chauvin, A., Dufresne, P., Oberlin, C. 2008. Archéologie du bâti et archéométrie au Mont-Saint-Michel, nouvelles approches de Notre-Dame-sous-Terre. Archéologie Médiévale 38, 71–122.

Tatton-Brown, T. 1980. St Martin's church in the 6th and 7th Centuries. In: Hobbs, J.E. and Sparks, M. (ed.). The Parish of St Martin and St Paul, Canterbury: Historical essays in Memory of James Hobbs. The Friends of St Martin, Canterbury, 12–18.

Webster, L. and Blackhouse, J. (eds.). 1991. The Making of England: Anglo-Saxon Art and Culture AD 600–900. The British Museum, London.

BRICK PRODUCTION AND BRICK BUILDING IN MEDIEVAL FLANDERS

Vincent Debonne

Abstract: In the past two decades, excavations, building archaeology and scientific dating techniques have generated new perspectives on brick architecture in the medieval county of Flanders. Due to the dearth of robust evidence, the pioneering role of the Cistercians in medieval brick production in Flanders is uncertain. Meanwhile, archaeological finds have demonstrated that production of roof tiles in Flanders dates back as far as the 10th century; however, its technical evolution into brick production remains unclear.

More than solely a cheap substitute for stone, the introduction and popularization of brick in medieval Flanders has to be understood within the dynamic building trade in a densely populated and urbanized county. Not coincidentally, the breakthrough of brick in the second half of the 13th century is simultaneous with innovations in stone building and roof construction. All are characterized by rationalization of the building trade, standardization of building materials and regularity in the execution of the architectural design.

Keywords: Brick, brick architecture, building ceramics, brick production, building archaeology, Flanders

Introduction

Awareness of the brick heritage of Flanders developed in the second half of the 19th century, a more scientific approach emerging after WW II with the rise of medieval archaeology and building archaeology. Since the late 20th century, urban archaeology and scientific dating techniques, mainly dendrochronology, have generated new perspectives on medieval brick production and brick building in Flanders. Drawing from both underground and building archaeology, this contribution firstly offers an overview of brick architecture in the county of Flanders in the 13th and 14th centuries. The second part addresses the 'lifecycle' of brick prior to the actual completion of the architectural design. Aspects regarding building typologies, formal architectural features and finish of the interior and exterior of medieval brick architecture are not discussed.

The geographic scope of our contribution is the medieval county of Flanders (Figure 1). Its boundaries were more or less fixed from 1191 onwards. It covered a territory corresponding to the present-day Belgian provinces of West and East Flanders, the southern bank of the Scheldt estuary now being part of the Netherlands and the western half of the French Département du Nord. A dense network of land routes and waterways connected medieval Flanders to international commercial traffic. As population growth accelerated, urbanization increased explosively; by 1300 Flanders was the most densely populated fief between Rhine and Loire.

Brick and the question of its origins

In a charter dated to 1223, Johanna of Constantinople, Countess of Flanders (1205–1244), confirms her bestowal to the Cistercian abbey of Boudelo of two hectares of land 'to make bricks'[1]. It is traditionally considered proof of brick production by Boudelo abbey, although stricto sensu the only conclusion that can be drawn from this is the gift of land containing clay deemed suitable for brick production. The Dunes, another Cistercian abbey, is generally considered to be the earliest brick-built complex in Flanders. This is based on the 15th-century abbey chronicle which mentions the start of the rebuilding of the abbey church in 1214[2]. However, apart from English stone acquired by abbot Elias (1189–1203), the chronicle does not mention any of the building materials[3]. The medieval wall fragments of The Dunes unearthed during the 20th century demonstrated an overall use of brick, but unfortunately these have been heavily restored and lost much of their material authenticity[4]. Due to a lack of excavations, the role of the Cistercian abbey of Ter Doest in the introduction of medieval brick in the Bruges area is unclear. The rebuilding of the abbey church is ascribed to abbot John (1243–53)[5]. However, during his abbacy, brick was already in use in Bruges.

In short, due to the dearth of robust evidence from both written and built sources, the pioneering role of the Cistercians in medieval brick production in Flanders is uncertain. Meanwhile, archaeological finds have demonstrated that the production of buildings ceramics in Flanders predates the supposed Cistercian invention of brick in the early 13th century. Roof tiles of the tegula and imbrex type were made as early as in the 10th century and continued to be produced into the early 13th century when the rectangular flat roof tile came into use. Tegula and imbrex roof tiles have been found on aristocratic and monastic (Benedictine) sites in the Scheldt and Dender

---

[1] [...] duo bonaria terre apud stekela ad conficiendes lateres. Ghent, State Archives, Boudelo n° 8, f° 166 r° & v°.
[2] As mentioned among others in Hollestelle 1961, 21–22 and Dewilde and De Meulemeester 2005, 189.
[3] Van de Putte and van de Casteele 1864, 35.
[4] Dewilde and De Meulemeester 2005.
[5] Van de Putte and Carton 1845, 12–13. The barn of the abbey's farmstead is all that remains of the medieval buildings of Ter Doest. Traditionally dated to c. 1275–1280 based on stylistic features, brick sizes and masonry bonds, the roof has been tree-ring dated to the second half of the 14th century (Nuytten 2005, 66–67).

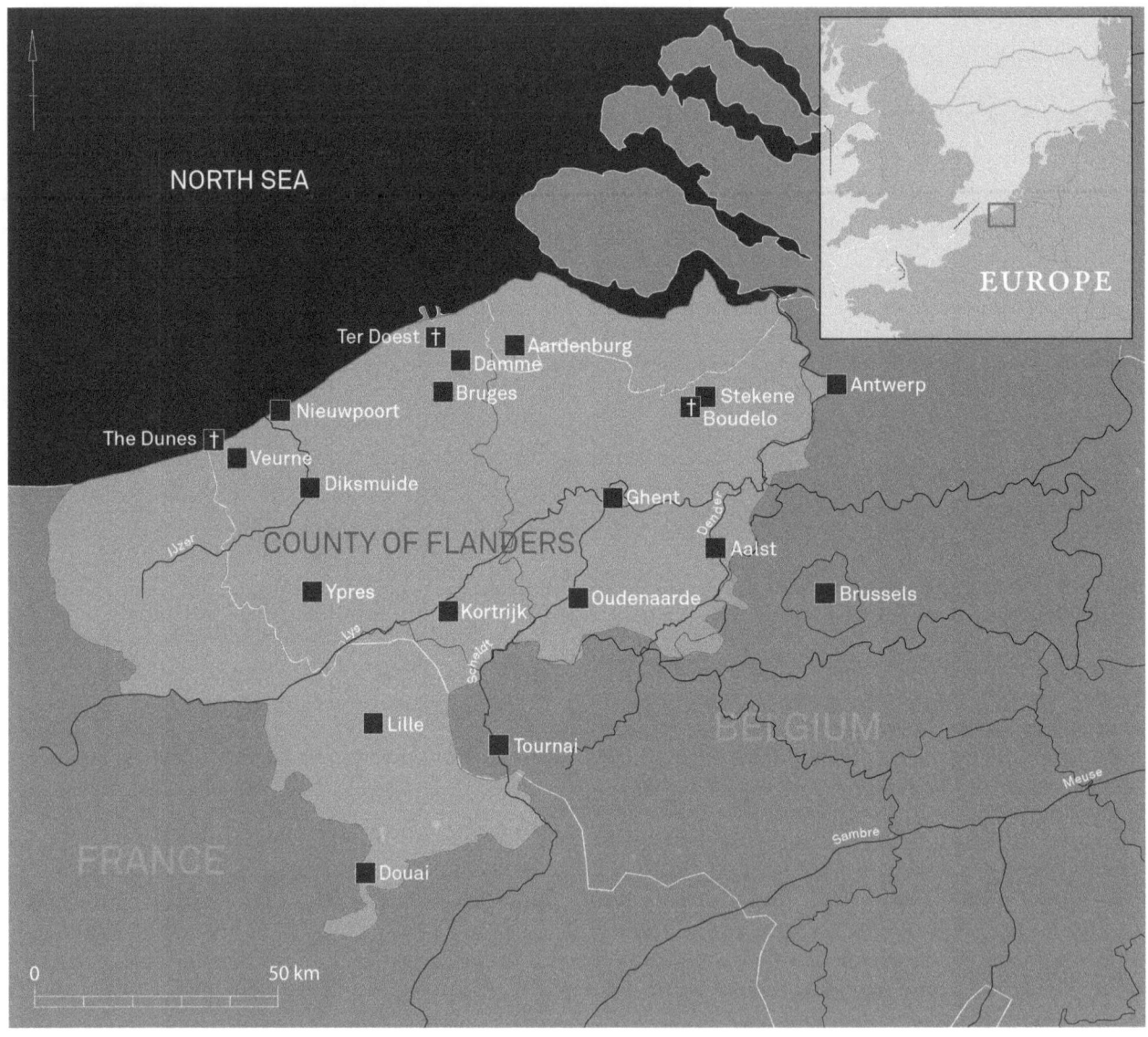

Figure 1. Map of the medieval county of Flanders (Nele Van Gemert, Flanders Heritage Agency).

valleys (Petegem, Ename, Dikkelvenne, Moorsel, Aalst)[6]. Presumably tegulae and imbrices were also produced and used elsewhere in Flanders – perhaps they have been found at excavations but were catalogued as Roman instead of medieval – most likely in the castella of the Counts of Flanders. According to the diary of the cleric Galbert, a detailed account of the events leading up to and following the murder of Count Charles the Good in 1127, the cupola of the church of St Donatian in the Count's domain in Bruges was built using building ceramics[7]. In Galbert's lifetime, the collegiate church of St Donatian was the 10th-century octagonal church modelled after the palatine chapel of Aachen[8]. Galbert narrates how the original wooden covering had been lost in a fire and replaced by a vault made of fire-resistant ollis et lateribus. 'Ollis' may have referred to acoustic vases, presumed fragments of which were found during excavations in 1987–88 close to the 10th-century foundations of the church[9]. 'Lateribus' probably indicates thin, tile-like bricks, easy to fit into the dome shape of a cupola. Probable fragments of these tegulae were likewise found during the above mentioned excavations[10]. Roof tiles were in use in Bruges at least by the early 13th century, as appears from a charter dating to 1232 for the nearby town of Aardenburg. In it, Count Ferrand (1212–33) decrees the covering of houses with tiles (lateribus), lead (plumbo) or wooden planks (asseribus) as customary in Bruges (secundum usus et consuetudines ville Brugensis)[11]. The medieval terms for building ceramics[12] bear testimony to the relationship between early bricks and roof tiles. In Latin

---

[6] Callebaut 1981, 15–18; Callebaut et al. 1987, 267; De Groote and Moens 1995, 138; Pieters et al. 1999, 135–141.
[7] Pirenne 1891, 60: Stabat autem ecclesia beati Donatiani aedificata in rotundum et altum, contecta fictitio opere, ollis et lateribus cacuminata [...]
[8] Devliegher 1991a, 46–63; Devliegher 1991b, 118–125. The 10th-century church was replaced by a new building at the end of the 12th century, which in turn was demolished in 1799 – 1800.
[9] Verhaeghe and Hillewaert 1991.
[10] Ibidem.
[11] Dezutter 1975.
[12] Hollestelle 1961, 49–53.

Figure 2. Bruges, St John's hospital, north wall of the middle ward. The arches of grey Tournai limestone were added during construction of the north ward (1266–80d).

texts, later may indicate either a brick or a roof tile. The Middle Dutch word tighel or tichel, derived from the Latin tegula, indicates both tiles and bricks. Appearing in the late 13th century, kareel (careel, quareel) solely refers to brick[13].

The seemingly sudden introduction of brick in the early 13th century probably resulted from a technical spin-off of existing tile production, in order to respond to new demands from the building industry in a society undergoing intense economic and demographic growth. In this environment, monastic communities were no longer the main patrons of architecture; construction progress was now also driven by the flourishing cities of the county.

The use of brick in the county of Flanders in the 13th and 14th centuries

The middle ward of St John's hospital in Bruges is the earliest extant brick building in Flanders (Figure 2)[14]. It is dated to the second quarter of the 13th century by tree-ring analysis of its impressive oak truss (1226−41d)[15], making it contemporary with the three lower storeys of the bell tower of Our Saviour's parish church (Figure 3)[16]. Dating to the first half of the 13th century, this brick-built tower has a facing of local fieldstone and volcanic tufa. Brick also quickly found acceptance among the Bruges citizenry, for example in the house known from medieval sources as 'ten Rooden Steene' ('the red stone house'), dating to the second quarter of the 13th century[17]. The chapter of the church of Our Lady on the other hand was hesitant to use brick as a fully-fledged building material. The nave (1240−50d) of this collegiate church, directly opposite St John's hospital, is built in Tournai limestone[18]. Brick was used only marginally to cover the exterior wall passage of the clerestory, impossible to see by passers-by in the

---

[13] For example in Dordrecht in the county of Holland (1285), Ghent (1288) and Bruges (1291) (Gysseling 1977, passim). The modern Dutch word baksteen ('baked stone') is first attested in Damme in 1465 (Haslinghuis 20055, 46).
[14] Esther 1976, 265–273.
[15] Dates obtained by tree-ring analysis are indicated by a 'd'. Five layers of brick in a wall of volcanic tuff north of the middle sick ward have been dated to the 12th century (Wets 2008, 152–157). Pending scientific dating, this date remains speculative.
[16] Devliegher 1981, 65–75.
[17] Van Eenhooge 2007, 25–30. The house name 'ten Rooden Steene' is first recorded in 1425.
[18] Devliegher 1954, 188–201; Devliegher 1997, 104–106; Van Eenhooge 2009, 25–28.

Figure 3. Bruges, Our Saviour's church, lower storeys of the bell tower.

Figure 4. Bruges, bell tower of the church of Our Lady.

street. The design and stone material of the nave were a means of the canons to distinguish themselves from other architectural patrons in Bruges, in particular the clergy of Our Saviour's church who claimed to be the oldest and most important ecclesiastical institution in the city[19].

As construction of a new choir of the church of Our Lady started in c. 1270–80, its chapter played a part in the overall rise in popularity of brick in Bruges during the second half of the 13th century. It is a period marked by vast brick building sites: the northern (1266–80d) and southern (1283–90d) wards of St John's hospital[20], that of the hospital of the Potterie (1276–96d)[21], the béguinage church (begun shortly after 1244)[22], the choir of Our Saviour's church (c. 1275–1300)[23] and the belfry and cloth hall (begun 1280)[24]. The most ambitious of them all, both in terms of size and technical prowess, was the bell tower of the church of Our Lady (Figure 4)[25]. After a laborious start around 1270, the tower was completed in c. 1350, being at the time the tallest brick building in the Low Countries. The medieval brick heritage of Bruges also consists of numerous remains of townhouses. Currently up to 170 houses dating to between 1200 and 1350 have been identified[26].

As Bruges developed into an international trading centre in the course of the 13th century, the surrounding area experienced formidable economic growth. As in Bruges, the second half of the 13th century marks the breakthrough of brick, coinciding with the prosperity of the trading towns along the Zwin, the natural channel connecting Bruges to the North Sea. In Damme, the parish church of Our Lady (c. 1225–50 to 1312–15d) (Figure 5) and the ward of St John's hospital (1270–85d) are reminders of the town's prosperity as the medieval outport of Bruges[27].

For medieval brick architecture in the Veurne region, reasonably reliable dates are available from the second quarter of the 13th century onwards[28]. While the brick remains of the Cistercian abbey of The Dunes have been heavily restored, the monumental remains of the brick barns at the former granges of Ten Bogaerde (Figure 6) and Allaertshuizen have escaped drastic interventions[29]. According to the 15th-century chronicle of The Dunes abbey, the barns were erected during the abbacy of Nicholas

---

[19] Devliegher 1981, 18 (n. 6).
[20] Esther 1976, 273–283.
[21] Tahon 2004.
[22] Devliegher 1954, 186–188.
[23] Devliegher 1981, 75–82.
[24] Devliegher 1965, 43–44.
[25] Devliegher 1997, 99, 107–108.

[26] Van Eenhooge 2007, 23–25.
[27] Devliegher 1971, 56–155; Debonne and Haneca 2011.
[28] Lehouck 2008.
[29] Ten Bogaerde lies at 2 km distance from the abbey in Koksijde, Allaertshuizen at 10 km in the commune of Wulpen.

Figure 5. Damme, church of Our Lady (Kris Vandevorst, Flanders Heritage Agency).

Figure 6. Koksijde, eastern front of the barn of Ten Bogaerde grange.

of Belle (1232−53)[30]. In Veurne, the choir (1265−75d) of the collegiate church of St Walburga[31] was under construction by the second quarter of the 13th century. Besides being the largest surviving medieval brick building in the region, it also ranks as one of the most accomplished Gothic churches in Flanders. Modelled after northern French churches, the choir features an ambulatory with radiating chapels, flying buttresses and quadripartite rib vaults. Simultaneous with the choir of St Walburga, a sturdy brick tower was built at the west end of the then Romanesque parish church of St Nicholas[32]. Remains in brick of the 13th-century 'landhuis', the administrative seat of the Veurne castellany, are conserved in the present-day 17th-century building. Remains of brick townhouses dating to the 13th and 14th centuries have been identified in the city centre[33], the most striking being cellars covered by ornate brick rib vaults resting on columns.

Unfortunately, much of the medieval heritage in southwestern Flanders was destroyed during 1914−18 as the First World War levelled the towns of Nieuwpoort, Diksmuide and Ypres. Excavations in Ypres, once the third largest city in medieval Flanders, suggest a popularization of brick in the second half of the 13th century[34]. For the construction of the west wing of the cloth hall, the biggest public building in medieval Flanders, the municipal accounts for the years 1285 to 1287 list an impressive number of bricks (2,045,350), roof tiles (284,000) and floor tiles (343,000)[35]. Bricks were bought from various suppliers as far as Diksmuide[36], some of whom also traded in stone and timber.

In Ghent, at the confluence of Scheldt and Lys, the first monumental use of brick is in the ward of the Bijloke hospital (1250−55d). The building has an exterior facing of Tournai limestone − transported to Ghent via the Scheldt − cladding a brick core[37]. The first quarter of the 14th century represents the peak of brick architecture in Ghent, the most notable architectural realizations being the east (1310−30d) and south wing (1318−38d) of Bijloke abbey, renowned for its brick gables with elaborate blind tracery (Figure 7), the Carmelite church (1320−30d) and the chapel of the wool weavers' hospice[38]. The chronology of brick in house building remains tentative. It seems that at least from the 14th century, brick was being used in house building.

The exterior look of many medieval buildings in Oudenaarde, halfway between Ghent and Tournai on the Scheldt, is marked by Tournai limestone[39]. Until recently little was known about medieval brick architecture in

Figure 7. Ghent, Bijloke abbey, western front of the refectory.

Oudenaarde since many buildings are brick-built structures hidden behind a stone facing. Such is the case for the cloth hall, built in the 1330s, and the church of the former Cistercian nunnery of Maagdendale (1292−1309d). Chance discoveries such as No 7 Hoogstraat, a well-preserved medieval house, suggest that brick came into use for house building in Oudenaarde during the second half of the 13th century.

The first reliably dated use of brick in Kortrijk dates from the 1290s[40]. Between 1297 and 1301, Philip IV, King of France (1285−1314), had a castle built in the Count's domain of Kortrijk. Excavations demonstrated that the castle was built in Tournai limestone and Artesian sandstone for the exterior facing and brick for the inner structure. We know from written sources that during the castle's construction the present-day choir of the nearby collegiate church of Our Lady was about to have its roof assembled. The choir has an exterior of large Tournai limestone ashlars hiding the brick core. Since the French castle was built very rapidly during the Franco-Flemish war of 1297−1305, one can assume that by then brick production in the Kortrijk area was already past its experimental stage. Brick also found its way into domestic architecture; around 1300 the wooden covering of several private cellars in Kortrijk was replaced with brick vaults. The most conspicuous of all

---

[30] Van de Putte and van de Casteele 1864, 9.
[31] Devliegher 1954, 319−327.
[32] Devliegher 1954, 316−319.
[33] Lehouck 2002.
[34] Dewilde 2008, 234−237.
[35] Des Marez and De Sagher 1909, 74−99. The medieval municipal accounts of Ypres were published in Idem, then destroyed in WW I.
[36] Idem, 77.
[37] Laleman and Stoops 2008, 164−166.
[38] Idem, 167−177.
[39] Debonne 2008, 189−191.

[40] Idem, 191−194.

medieval brick buildings in Kortrijk is the 14th-century belfry in the central market square.

Besides urbanization, the intense demographic growth of the county led to the exploitation and population of wastelands. The occupation of the large stretches of uncultivated land between Bruges, Ghent and Antwerp is evidenced by the foundation of new parishes in the 13th century, which explains the striking density of 13th- and 14th-century brick churches in this area. Medieval brick production in the Waasland, a historic region between Ghent and Antwerp, was centred in and around Stekene, the village where the clay grounds 'to make bricks' could be found that were given to Boudelo abbey in 1223 by the Countess of Flanders. The first explicit mention of brick production in the Waasland is to be found in a book of rents from Boudelo abbey dated to 1261–63, listing two generations of brickmakers and their brick yard in Stekene[41]. Excavations have shown the abbey of Boudelo, which is no longer extant, to have been built using brick made from local Rupelian clay[42]. Several parish churches in the immediate area of the abbey were either rebuilt or newly built in brick during the second half of the 13th century. Brick production in Stekene occurred on an increasingly industrial scale from the 14th century onwards, with exports to Ghent and even as far as Kortrijk[43].

By 1300, brick was a well-established building material in most of Flanders, although some regions lagged behind. Brick building in the castellanies of Lille and Douai does not seem to have taken off until the 14th century. Built simultaneously with the castle in Kortrijk by King Philip IV, the 'Château de Courtrai' in Lille (1298–1302) was constructed solely using stone[44]. The first written mention of brick in Lille dates from 1318 and relates to the construction of the city walls[45]. In Douai, the use of brick is first recorded in 1376, also relating to the city's fortifications[46]. The introduction of brick might well have been earlier, but confirmation provided by (building) archaeology is lacking. The later introduction of brick in Lille and Douai as compared to the rest of Flanders lies partly in the presence of quarries at only a few kilometres outside the city gates. For the region between Scheldt and Dender, the relatively late introduction of brick is only partially tied to the presence of easily workable limestone. Here, urbanization was not as intense as in the regions west of the Scheldt (all cities are located along the Dender), resulting in a less dynamic development of the construction industry. Even in the city of Aalst, which underwent considerable growth in the 13th century, brick was unknown until the great fire of 1360. Before the fire, local limestone was the preferred material for major religious and public buildings such as the aldermen's house, whereas ordinary houses were half-timbered.

[41] Ghent, State Archives, Boudelo 552, f° 2r°.
[42] De Belie 1997.
[43] Thiron 2010.
[44] Blieck 1997, 190–196.
[45] Blieck 1988, 113–114.
[46] Salamagne 2001, 164.

Figure 8. Snellegem parish church (12th century), masonry in glauconitic sandstone.

From clay pit to building site

Primary materials

The dense network of navigable waterways facilitated the transport of stone quarried on the southern fringes of the county to the north. Tournai limestone had been used in the valleys of Scheldt and Lys since Roman times; limestone from the Artesian hills south of the Aa was exported to the IJzer basin and up to the middle course of the Lys. Stone types indigenous to the county of Flanders are greyish-green glauconitic sandstone (Figure 8) found between Bruges and Ghent as fieldstone, and brown ferruginous sandstone from the hills south of Ypres and Oudenaarde (Figure 9)[47]. However, both stone types are difficult to exploit and, especially in the case of the irregular fieldstone, hardly suited for regular masonry, resulting in massive, thick-walled constructions.

Brick on the other hand was easy to produce in standardized shapes and sizes, which allowed for regular, structurally coherent masonry and less massive walls, thus reducing the amount of material needed. Apart from the polder clay in the West Flemish coastal plain, clays suitable for brick production are the alluvial deposits of Scheldt and Lys and the Ypresian and Rupelian clays in the Kortrijk-Ypres area

[47] Dusar et al. 2009, 255–262, 503–509.

Figure 9. Wulveringem parish church (12th century), masonry in ferruginous sandstone.

Figure 10. Sinaai parish church (second half of the 13th century), header showing pressing folds.

and the Waasland respectively[48]. As for combustibles for brick production, medieval exploitation of peat has been demonstrated in the coastal plain, the Waasland and north of Ghent[49]. Export of peat from these areas to the south of the county may have begun as early as c. 1200[50].

The identity of those in charge of clay exploitation is largely unknown. Historiography has traditionally stressed the importance of the Cistercians, leaving underexposed the possibly substantial role of the nobility. According to feudal custom, the count of Flanders was the prime owner of the waste grounds in the county. As suggested by the Countess' bestowal in 1223 of land to Boudelo abbey 'to make bricks', a considerable amount of land with clay deposits may have been the property of the Count and his vassals. The popularization of brick from the second half of the 13th century onwards probably prompted landowners to put their lands to lucrative use by producing and selling bricks. In 1321, the lord of Schelderode, 13 km south of Ghent, organised the transport of bricks made at his own brickyard along the river Scheldt[51]. Clay exploitation and brick production were usually given in rent or as a lease. According to the Boudelo book of rents of 1261–63, the abbey's brick yards were managed in rent by a lay brickmaker – not a lay brother as is often assumed for Cistercian brick production. The brickyards in Ramskapelle owned by the city of Bruges in the 1330s were leased to the operators of the brick kilns[52].

Brick production

Moulding

No medieval wooden moulding frames have survived but their use can be deduced from particular traces on bricks. The folds formed when clay was pressed into the moulding frame are easy to recognize on the headers (Figure 10). Rims on the upper side of bricks are burrs of clay, formed as the unfired brick was pushed out of the moulding frame (Figure 11). Sunken rims are the imprints of the moulding frame as it was pushed away from the sticky unfired brick (Figure 12). Intentional marks applied before firing are not uncommon, such as scrapes and indentations (Figure 13). Their significance is unknown.

Profiled bricks were produced as early as the second quarter of the 13th century in the Bruges and Veurne regions and from the second half of the 13th century elsewhere in the county as well. The most common types are those with chamfered edges and roll-mouldings. The largest variety in profiled bricks in Flanders has been found at the sites of the Cistercian abbeys of The Dunes and Boudelo[53]. The smooth surface (Figure 11) of the profiled sides indicates a moulding before firing with a wooden frame or a pliable metal thread, possibly after the basic rectangular brick shape had been created with a standard moulding frame. Parallel surface striations are traces of knives (Figure 14) used to cut the unfired brick into shape. The softly incised lines on the upper side of several moulded bricks from The Dunes and Boudelo were applied before firing and served to facilitate the precise execution of mouldings (Figure 15). Other marks applied before firing may be identified

---

[48] Gullentops and Wouters 1996, 36–48.
[49] Sosson 1977, 68; Augustyn and Thoen 1987, 99–107; Augustyn 1999, 27–43.
[50] Deforce et al. 2006, 147–149.
[51] Lachaert 1998, 104.
[52] Sosson 1977, 71–79.
[53] Debonne 2009.

Figure 11. Profiled brick (13th century) from Boudelo abbey, displaying clay burrs at the edges.

Figure 13. Sinaai parish church (second half of the 13th century), brick with indentations.

Figure 12. Brick from the Ten Bogaerde barn (second quarter of the 13th century), displaying sunken margins (Hans Denis, Flanders Heritage Agency).

Figure 14. Profiled brick from The Dunes abbey (13th century) with cutting marks.

as brickmaker's marks, similar to the marks left on stone blocks by stonecutters. The Roman numerals on bricks from The Dunes served as layout marks for masons at the building site (Figure 15). The different lines and marks on bricks from The Dunes and Boudelo are identical to those pictured on folios 21 and 32 in the sketchbook of Villard de Honnecourt (c. 1235), showing building parts and templates of profiled stone blocks of Reims Cathedral. Technical exchange between stonecutters and brickmakers is very likely in the case of The Dunes from where numerous profiled stone blocks have been collected, some of which are the exact counterparts of profiled bricks. A clear combination of profiled bricks and stone pieces can be seen in a contemporary of The Dunes abbey, the choir of St Walburga in nearby Veurne, built in c. 1225 – 1275 (Figure 16).

Figure 15. Profiled brick from The Dunes abbey (13th century), displaying incised lines and a stamped Roman numeral II.

Figure 16. Veurne, church of St Walburga. Column made of grey Tournai limestone, springing of the rib vault in pale yellowish limestone and brick for the remainder of the vault.

Brick sizes

As they decrease over time, brick sizes are often used as dating devices in (building) archaeology. Their chronological value however is far from absolute. As has been remarked[54], bricks of different sizes were used for different parts of a building. In the béguinage church of Bruges, the partition arches of the nave are built using bricks that are much wider (28 x 23 x 6.5cm)[55] than the ones used for continuous masonry (30 x 14.5–15 x 6.5–7cm). Bricks of different sizes were used in the vaults of St Walburga in Veurne: large ones (28.5 x … x 8.5cm) for the cresting, thinner ones for the springing points (… x 12.5 x 6.5cm) and smaller ones for the groins (26 x 12.5 x …cm). The building accounts of the Ypres cloth hall (1285–87) distinguish between tieules de tere (bricks), tieules de tere grans (large bricks) and tieules de tere petis (small bricks)[56].

In addition, one must take into account periods of rapid decrease in brick size when bricks of different sizes were simultaneously available. The reduction of brick sizes was particularly rapid in the late 13th and early 14th century; as the amount of brick makers grew so did the number of bricks of different sizes. The length of bricks in the church of Maagdendale in Oudenaarde, built in a short span of time around 1300, varies from 29 cm down to 26.5 cm. The Bruges area saw a considerable decrease in brick size, from 28.9 x 13.9 x 6.2cm in the years 1280 (Damme, church of Our Lady) over 24.5 x 11.5 x 5.25cm in the second quarter of the 14th century (Bruges, tower of Our Lady) to 21 x 9.5 x 5cm around 1350 (Bruges, northernmost side aisle of Our Lady).

The increase in brick production in the course of the 14th century urged councils of castellanies and cities, for example in Bruges and Ghent, to set a limit on the number of fixed brick sizes[57]. Building accounts show that bricks were paid by the numbers, not by the total weight of the charge. Brick merchants may have favoured smaller bricks whereas patrons probably preferred larger ones. Regulations on brick sizes may therefore have been attempts by local governments to reconcile the interests of both producer and consumer. Brick sizes in medieval Flanders, it should be added, did not uniformly coincide with the foot measure which, moreover, differed from castellany to castellany. Quite remarkably, the length of bricks in the barn of Ten Bogaerde, a grange of The Dunes abbey, corresponds to the Roman foot (29.6cm).

Apart from possible chronological values, brick sizes may allow regional traditions in brick production to be distinguished. For example, the traditional assumption of a 'typical' Cistercian brick size is hard to uphold when comparing brick sizes from the abbeys of The Dunes, Ter Doest and Boudelo. The presence of c. 30cm long bricks

---

[54] Lehouck 2008, 229.
[55] Brick sizes are noted as length x width x height. '…' refers to the non-measurable sizes of bricks when in masonry.
[56] Des Marez and De Sagher 1909, 76, 82, 86.
[57] Sosson 1977, 78–79; Laleman and Stoops 2008, 175.

commonly known as 'kloostermoppen' (monastic bricks) in the remains of the abbeys and their granges is seen as proof of The Dunes exporting its brick technology to other abbeys. However, bricks in the barn of Ter Doest have a different size (32 x 14.8 x 7.6cm) from those of the barn of the Ten Bogaerde grange (29.4 x 13.2 x 8cm). On the other hand, bricks with a similar size as those at Ter Doest occur in buildings in and around Bruges – the brick sizes at Ter Doest are not so much Cistercian but local. The bricks of Boudelo abbey (29.8 x 14.1 x 6.6cm) resemble those of The Dunes in terms of their length, but they are markedly thinner than in The Dunes. The reduced height of bricks as compared to their length and width is characteristic of the Waasland, where by the 14th century bricks of 4 to 4.5cm thickness were in use, thinner than bricks elsewhere in Flanders at the time.

Drying, firing and transporting bricks

The zigzag stacking of unfired bricks in drying sheds, as depicted in late medieval iconographic sources, was customary until the automation of brick production in the early 20th century. Remains of drying sheds have not yet been found archaeologically, but they are known from written sources. In the building accounts of the castle built in Courtrai between 1394 and 1404 by Philip the Bold, Duke of Burgundy and Count of Flanders (1384–1404), the brick-drying shed is referred to as a haghenhuus ('hedge house'). Between the stakes supporting the roof of the shed, hedges were placed to shelter the unfired bricks from heavy precipitation and simultaneously enabling air circulation to facilitate the drying process.

The construction of the pre-industrial brick kilns so far excavated in Flanders is consistently the same[58]. Unfired bricks were stacked in cube-shaped pallets with fire channels for combustibles (charcoal, peat) at the bottom. The earliest known excavated brick kilns are located near Ramskapelle, northwest of Bruges[59]. Dating to the 14th century, they were part of the local brick industry supplying the city of Bruges. The use of the aforementioned kilns can be deduced from differences in surface texture of bricks. Glazed surfaces indicate partial overexposure to heat due to corbeled stacking of bricks in the kiln, whereas the common orange-red colour indicates surfaces were sheltered from overheating by superimposed bricks (Figure 17). Such glazed surfaces, as traces of stacking in kilns, are already present on 13th-century bricks. Colourfully glazed bricks such as in northern Germany are unknown in medieval Flanders. The use of bricks with (accidentally) dark glazed headers became increasingly popular in the 15th century for the design of masonry marks. Additional working of bricks after firing has so far only been observed at the sites of The Dunes and Boudelo. Some profiled bricks show evidence of working with a claw tool, presumably used for the final flourish once the bricks had been assembled into composite piers, shafts etc.

[58] Hartoch 2010.
[59] Patrouille 2002.

Figure 17. Kloosterzande (Hulst, the Netherlands), partly glazed brick in the chapel of Hof te Zande grange (second quarter of the 13th century).

Brickyards were located either close to the building sites or a navigable waterway. In Ghent, the placename Tichelrei indicates the former presence of a medieval brickyard, both safely and conveniently located just outside the late 12th-century city moat. In Kortrijk, the bricks used in 1392–93 for a tower of the city wall were brought to the building site by ship via the river Lys. The bricks produced in villages northwest of Bruges (Dudzele, Lissewege, Ramskapelle) were easy to transport via the dense regional network of waterways. The rise of Stekene as one of the main brick producing villages in Flanders was partly thanks to its location near navigable waterways connecting it to Ghent and, from there on, to Bruges and the Scheldt and Lys basins.

Masonry bonds

The early evolution of brick masonry can be traced to Bruges, where brick was initially used to encase rubble walling, alternated with layers of volcanic tufa. This is a building technique that does not necessarily require a regular masonry bond; without the need to distinguish between headers and stretchers, bricks were cut to the appropriate length instead, as if they were stone blocks. Evidence for the cutting of bricks can be observed in the middle ward of St John's hospital and the bell tower of Our Saviour's church, where bricks have varying lengths from 10 to 33.7 cm (St John's hospital) and 20 to 34 cm (Our Saviour's church).

Such contrasting interplay of red brick and white stone (Figure 2), in casu tufa, derives from Roman architecture. In the case of Bruges, presumably the link with Roman architecture was not direct, but rather a continuation of building traditions of the Carolingian era, perhaps imported into Bruges together with the export of volcanic tufa from the Rhineland. Contrasting layers of brick and stone mark

Figure 18. Koksijde, western front of the barn of Ten Bogaerde grange. Buttress with queen closers, continuous masonry in Flemish bond.

Figure 19. Damme, St John's hospital. Eastern front in English bond.

several Ottonian churches in Cologne, a conscious reference to the city's Roman past[60].

In the upper storeys of St John's hospital and the bell tower of Our Saviour's church, the combined use of brick and tufa is abandoned in favour of continuous brick masonry. The brick masonry in the ward of St John's hospital is irregular at first, showing an improvized alternation of stretchers and headers. In the parapet and the western gable of the ward, Flemish bond[61] comes into use. The use of a regular masonry bond allowed for constructively coherent buildings and saving on building materials. Whereas the use of fieldstone resulted in massive, thick-walled buildings with irregular masonry, brick enabled carefully planned constructions with buttresses and vast sparing arches. Ambitious buildings programmes were now undertaken, for example the bell tower of Our Saviour's church, built in the first half of the 13th century. Until the construction of the belfry (1280), the 35m high tower was the tallest building in Bruges[62]. In contrast to Frisia and Northern-Germany, double stretcher-header bonds (Scotch bond) were unknown in medieval Flanders.[63]

The remains of the barns of Ten Bogaerde and Allaertshuizen, built in the second quarter of the 13th century, show a complete mastery of the Flemish bond with queen and king closers used systematically at the angles of buttresses and jambs (Figure 18). As in the tower of Our Saviour's church in Bruges, the use of a regular masonry bond enabled a structural articulation of the building, in the case of the barns structuring the front elevations by a succession of pointed sparing arches. By the second half of the 13th century the Flemish bond was in use in most of Flanders, in the Bruges and Veurne area in the west as well as in Ghent, Kortrijk, Oudenaarde and the Waasland in the east. The last applications of Flemish bond date from the first quarter of the 14th century. English bond (Figure 19) was introduced as early as the 1270s (e.g. the church of St Walburga in Veurne, 1265–75d, and St John's hospital in Damme, 1270–85d) and commonly in use by the early 14th century. How and when English bond developed into English cross bond is poorly known; the latter was common by at least the 16th century and remained so into the early 20th century.

---

[60] Beuckers et al. 2002, 240–245.
[61] Flemish bond consists of regularly alternating headers and stretchers.
[62] The height of the tower was raised in the 14th and 15th centuries, and it was crowned by a neo-romanesque spire in 1865 – 1872 (Devliegher 1981, 65).

[63] Scotch bond has only been observed sporadically in Bruges within masonry of irregular or Flemish bond (Wets 2008, 150)

Conclusions

Scientific dating techniques have enabled a fresh look at medieval brick architecture in Flanders. As an increasing number of buildings are now accurately dated, it appears that the explosive rise in the popularity of brick from the mid-13th century onwards coincided with innovations in other branches of the building industry. In stone construction, the shift from irregular rubble masonry to regular ashlar masonry was contemporary with the appearance around 1250 of regular brick masonry bonds. Roof construction during the 13th century was similarly innovative, as the use of collar purlins and inclined principal trusses enabled the covering of spaces much wider and higher than before. Contrary to common assumption, brick in medieval Flanders was more than a cheap substitute for stone; indeed it typified the standardization and rationalization of the late medieval construction industry.

New questions on medieval brick production involve physical and technical aspects, such as the evolution from (post)-Carolingian roof tile production to brick production and regional tendencies in brick production. Provenancing of stones by means of petrographic analysis is relatively common in research into medieval architecture, whereas the long-distance trade in bricks in medieval Flanders, well known from written sources, still awaits material confirmation. Scientific research into building archaeology needs not to be limited to dating; examination of the physical properties of bricks is also likely to generate new insights into medieval brick production.

References

Augustyn, B. and Thoen, E. 1987. Van veen tot bos. Krachtlijnen van de landschapsevolutie van het Noordvlaamse Meetjesland van de 12e tot de 19e eeuw. Historisch Geografisch Tijdschrift 5, 97–112.

Augustyn, B. 1999. De veenontginning (12de-16de eeuw). Geschiedenis van volk en land van Beveren 5. Beveren, Gemeentebestuur Beveren.

Beuckers, K. G., Cramer, J. and Imhof, M. 2002. Die Ottoner. Kunst – Architektur – Geschichte. Petersberg, Michael Imhof Verlag.

Blieck, G. 1988. Recherches sur les fortifications de Lille au Moyen Age. Revue du Nord 70/276, 107–122.

Blieck, G. 1997. Le château dit de Courtrai à Lille de 1298 à 1339: une citadelle avant l'heure. Bulletin Monumental 155/3, 185–206.

Callebaut, D. 1981. Het oud kasteel te Petegem. De Karolingische curtis en haar ontwikkeling tot de XIIde eeuw, Archaeologia Belgica 237. Brussels, Nationale Dienst voor Opgravingen.

Callebaut, D., Pieters, M. and Van Durme, L. 1987. De Sint-Pietersabdij te Dikkelvenne (gem. Gavere), Archaeologia Belgica 3, 265–268.

De Belie, A. 1997. De Boudelo-abdij archeologisch onderzocht. Sint-Niklaas, Culturele Kring Boudelo.

De Groote, K. and Moens, J. 1995. De oudste stadsversterking van Aalst (prov. Oost-Vlaanderen), Archeologie in Vlaanderen 4, 95–148.

Des Marez, G. and De Sagher, E. (eds) 1909. Comptes de la ville d'Ypres de 1267 à 1329. Brussels, Kiessling.

Debonne, V. and Lachaert, P.-J. 2007. Een middeleeuws huis in Oudenaarde (prov. Oost-Vlaanderen): historisch en bouwhistorisch onderzoek in Hoogstraat 7. Relicta. Archeologie, Monumenten- en Landschapsonderzoek in Vlaanderen 3, 237–288.

Debonne, V. 2008. Bouwen met baksteen in het Kortrijkse en het Oudenaardse tijdens de middeleeuwen. In Coomans, Th. and Van Royen, H. (eds), Medieval Brick Architecture in Flanders and Northern Europe: The Question of the Cistercian Origin. Novi Monasterii 7, 185–202.

Debonne, V. 2009. Production of Moulded Bricks on a Gothic Building Site. The Case of the Thirteenth-Century Abbeys of the Dunes and Boudelo (Belgium). In Kurrer, K.-E., Lorenz, W. and Wetzk, V. (eds), Proceedings of the Third International Congress on Construction History (Brandenburg University of Technology Cottbus, Germany, 20th-24th May 2009), 459–464. Cottbus, Chair of construction history and structural preservation Brandenburg University of Technology.

Debonne, V. and Haneca, K. 2011. Baksteen en boomringen: een verfijnde bouwchronologie van het hallenkoor van de Onze-Lieve-Vrouwkerk in Damme (prov. West-Vlaanderen). Relicta. Archeologie, Monumenten- en Landschapsonderzoek in Vlaanderen 7, 67–100.

Deforce, K., Bastiaens, J. and Ameels, V. 2006. Archeobotanisch bewijs voor ontginning en lange-afstandstransport van turf in Vlaanderen rond 1200 AD: heropgegraven veen uit de abdij van Ename (Oudenaarde, prov. Oost-Vlaanderen). Relicta. Archeologie, Monumenten- en Landschapsonderzoek in Vlaanderen 1, 141–154.

Devliegher, L. 1954. De opkomst van de kerkelijke gotische bouwkunst in West-Vlaanderen gedurende de XIIIe eeuw. Bulletin van de Koninklijke Commissie voor Monumenten & Landschappen 5, 179–345.

Devliegher, L. 1965. Beeld van het kunstbezit. Kunstpatrimonium van West-Vlaanderen 1. Tielt – Den Haag, Lannoo.

Devliegher, L. 1971; Damme. Kunstpatrimonium van West-Vlaanderen 5. Tielt – Utrecht, Lannoo.

Devliegher, Luc 1981. De Sint-Salvatorskatedraal te Brugge. Geschiedenis en architektuur. Kunstpatrimonium van West-Vlaanderen 7. Tielt – Bussum, Lannoo.

Devliegher, L. 1991a. De voorromaanse Sint-Donaaskerk en de romaanse westelijke kloostervleugel (onderzoek 1955 en later). In De Witte, H. (red.), De Brugse Burg. Van grafelijke versterking tot moderne stadskern. Archeo-Brugge 2, 46–92. Bruges, v.z.w. Archeo-Brugge.

Devliegher, L. 1991b. De voorromaanse, romaanse en gotische Sint-Donaaskerk. Evolutie en invloeden. In De Witte, H. (red.), De Brugse Burg. Van grafelijke versterking tot moderne stadskern, Archeo-Brugge 2, 118–136. Bruges, v.z.w. Archeo-Brugge.

Devliegher, L. 1997. De bouwgeschiedenis van de kerk. In de Smaele, H., Flour, E., Heyndrikx, V. and De Jonge, M. (reds), De Onze-Lieve-Vrouwekerk te Brugge, 99–108. Brugge, Jong Kristen Onthaal voor Toerisme.

Dewilde, M. and De Meulemeester, J. 2005. Archeologie, geschiedenis en bouwhistorie. In Vanclooster, D. (red.), De Duinenabdij van Koksijde. Cisterciënzers in de Lage Landen, 180–195. Tielt, Lannoo.

Dewilde, M. 2008. Bouwen met baksteen in middeleeuws Ieper. In Coomans, Th. and Van Royen, H. (eds), Medieval Brick Architecture in Flanders and Northern Europe: The Question of the Cistercian Origin. Novi Monasterii 7, 233–241.

Dezutter, W. 1975. Een stedelijk bouwvoorschrift uit 1232 tegen brandgevaar (Aardenburg en Brugge). In Dezutter, W. and Goetinck, M., Tentoonstelling op en om de bouwwerf. Ambachtswezen – oud gereedschap, 59–68. Bruges, Stedelijke Musea en Museum voor Volkskunde.

Dusar, M., Dreesen, R. and De Naeyer, A. 2009. Renovatie & restauratie. Natuursteen in Vlaanderen, versteend verleden. Mechelen, Kluwer.

Esther, J.-P. 1976. Monumentenbeschrijving en bouwgeschiedenis. In Sint-Janshospitaal Brugge. 1188/1976, 259–339. Bruges, Commissie van Openbare Onderstand.

Gullentops, F. and Wouters, L. 1996. Delfstoffen in Vlaanderen. Brussel, Ministerie van de Vlaamse Gemeenschap, Departement EWBL.

Gysseling, M. (ed.) 1977. Corpus van Middelnederlandse teksten (tot en met het jaar 1300). Reeks I: Ambtelijke bescheiden. The Hague, Nijhoff.

Hartoch, E. 2010. Archeologisch onderzoek naar baksteenovens in Vlaanderen: een overzicht. In Oost, T. and Van de Voorde, E. (eds), In vuur en vlam! Omgaan met baksteenerfgoed in Vlaanderen. Jaarboek voor Geschiedenis en Volkskunde. Monografie 1, 62–134. Antwerp, Provinciebestuur van Antwerpen.

Haslinghuis, E.J. 2005. Bouwkundige termen. Verklarend woordenboek van de westerse architectuur- en bouwhistorie. Leiden, Primavera Pers.

Hollestelle, J. 1961. De steenbakkerij in de Nederlanden tot omstreeks 1560. Arnhem, Van Gorcum.

Lachaert, P.-J. 1998. De landschapsgeschiedenis van Schelderode. Ruimtelijke ordening en organisatie van een cultuurlandschap (12de-18de eeuw). Unpublished master's thesis, University of Ghent.

Laleman, M. C. and Stoops, G. 2008. Baksteengebruik in Vlaamse steden: Gent in de middeleeuwen. In Coomans, Th. and Van Royen, H. (eds), Medieval Brick Architecture in Flanders and Northern Europe: The Question of the Cistercian Origin. Novi Monasterii 7, 163–183.

Lehouck, A. 2002. Het verstedelijkingsproces en de oudste stenen burgerwoning van Veurne. Handelingen der Maatschappij voor Geschiedenis en Oudheidkunde te Gent 56, 1–34.

Lehouck, A. 2008. Gebruik en productie van baksteen in de regio Veurne van circa 1200 tot circa 1550. In Coomans, Th. and Van Royen, H. (eds), Medieval Brick Architecture in Flanders and Northern Europe: The Question of the Cistercian Origin. Novi Monasterii 7: 203–232.

Nuytten, D. 2005. Bouwhistorisch onderzoek van de voormalige abdijschuur van Ter Doest. Bulletin van de Koninklijke Nederlandse Oudheidkundige Bond 104/2–3, 58–74.

Patrouille, Els 2002. Laatmiddeleeuwse baksteenindustrie te Zeebrugge (prov. West-Vlaanderen). Archeologie in Vlaanderen 6, 243–260.

Pieters, M., De Groote, K., Ervynck, A. and Callebaut, D. 1999. Tussen kapel en kerk: een archeologische kijk op de evolutie van de dorpskern van Moorsel (10de-20ste eeuw). Archeologie in Vlaanderen 5, 131–157.

Pirenne, H. (ed.) 1891. Histoire du meurtre de Charles le Bon comte de Flandre (1127–1128) par Galbert de Bruges, suivie de poésies latines contemporaines. Collection de textes pour servir à l'étude et à l'enseignement de l'histoire 10. Paris, Picard.

Salamagne, Alain 2001. Construire au Moyen Age. Les chantiers de fortification de Douai. Villeneuve-d'Ascq, Presses universitaires du Septentrion.

Sosson, Jean-Pierre 1977. Les travaux publics de la ville de Bruges XIVe-XVe siècles. Les matériaux, les hommes. Collection Histoire Pro Civitate 48. Brussels, Crédit Communal de Belgique.

Tahon, E. 2004. Brugge, Onze-Lieve-Vrouw ter Potterie. In Buyle, M. and Dehaeck, S. (reds), Architectuur van de Belgische hospitalen, M and L Cahier 10, 104–108. Brussel, Ministerie van de Vlaamse Gemeenschap.

Thiron, J. 2010. Stekene, meer dan 800 jaar productie van bouwkeramiek in een Vlaams dorp. In Oost, T. and Van de Voorde, E. (eds), In vuur en vlam! Omgaan met baksteenerfgoed in Vlaanderen. Jaarboek voor Geschiedenis en Volkskunde. Monografie 1, 41–61. Antwerpen, Provinciebestuur van Antwerpen.

Van de Putte, F. and Carton, C.-L. 1845. Recueil de l'abbaye de Ter Doest, Recueil de Chroniques, Chartes et autres Documents concernant l'histoire et les Antiquités de la Flandre-Occidentale. Première série: Chroniques des monastères de Flandre. Bruges, Vandecasteele-Werbrouck.

Van de Putte, F. and van de Casteele, D. 1864. Cronica et cartularium monasterii de Dunis, Recueil de Chroniques, Chartes et autres Documents concernant l'histoire et les Antiquités de la Flandre-Occidentale. Première série: Chroniques des monastères de Flandre. Bruges, Vandecasteele-Werbrouck.

Van Eenhooge, D. 2007. Middeleeuwse Brugse huizen en hofsteden langs de Spiegelrei. M&L. Monumenten, landschappen en archeologie 28/2, 21–45.

Van Eenhooge, D. 2009. De middeleeuwse sporenkappen van de Onze-Lieve-Vrouwekerk te Brugge. M&L. Monumenten, landschappen en archeologie 26/5, 22–44.

Verhaeghe, F. and Hillewaert, B. 1991. Bouwpotten in de oude Burgkerk. In De Witte, H. (red.), De Brugse Burg. Van grafelijke versterking tot moderne stadskern, Archeo-Brugge 2, 182–193. Bruges, v.z.w. Archeo-Brugge.

Wets, Th. 2008. Vroeg baksteengebruik in Brugge. In Coomans, Th. and Van Royen, H. (eds), Medieval Brick Architecture in Flanders and Northern Europe: The Question of the Cistercian Origin. Novi Monasterii 7, 147–162.

Figures

All figures and photographs by the author, except where noted otherwise.

Approach on Brick and its Use in Brussels from the 14th to the 18th Century

Philippe Sosnowska[1]

Abstract: Omnipresent in the citylandscape of Brussels, the brick of the Ancien Regime still remains an unknown materiel for the public and for the scientific community. New research conducted by the Research Centre for Archaeology and Heritage of the Université libre de Bruxelles provides a new focus on the fabrication and the use of this material in the city and its periphery. According to the historical and archaeological sources, the first witness of brick dates from the second half of the 13th century or from the beginning of the 14th century. From that moment onwards, the use continuously increased. Between the 14th and 16th centuries, the use of brick is contemporary with other material types such as stone and timber. From the 17th century onwards, brick constructions will supplant all other types of architecture. Changes in the production and the implementation of bricks are observed almost simultaneously, showing the innovative process made on ceramics and the capacity of masons to adapt their work to it. Research shows that at the end of the 17th century, and during the following century, the importation of 'foreign' bricks is more and more common. They will be used in mass, supplant the local product and provide deep changes in the way of brusselse brick production.

Keywords: architectural ceramics, architectural changes, bricks, brick bonds, Brussels, building material, construction

Introduction[2]

The use of brick in Brussels (former duchy of Brabant) and its periphery is a late phenomenon in relation to its appearance in the Northern Low Countries (currently Holland) and the county of Flanders. In the latter case, the first evidence dates back to the first third of the 13th century[3], the use of brick in the regional context studied so far is not attested before the 14th century, or perhaps in the second half of the 13th century.

Generally in our regions, brick and architectural ceramics remain the poor parent of research on building materials. A history of the production of architectural ceramics in Brussels, and in general in the whole duchy of Brabant, remains to be written. When considering paving tiles, roof tiles, or even bricks, the lack of synthesis and even the relative lack of information on the subject should be noted. Valuable informations, when they exist, are to be found in case studies. Nonetheless, it concerns three basic materials in the realm of construction in the studied region. The replacement of pavements and roofing tiles over the years has consequently removed the oldest traces of these materials from the built landscape, which could explain a certain lack of interest by historians, building historians, art historians and archaeologists. On the other hand, what to say about the brick of the Ancien Regime, whose use is so characteristic of this architecture that it is commonly called «traditional»[4], and who even transcends all architectural styles in our region[5]?

Without pretending to be exhaustive, this article aims to provide a first overview of the built landscape of Brussels sketched throughout the use of brick from the 14th until the 18th century. This will be an overview and a perspective of the data published to date as well as the results of our research[6].

The first evidence of the use of brick in Brussels (second half of the 13th century?–14th century), the Porte de Hal, built during the second half of the fourteenth century, is the sole remnant of the second defensive system that surrounded the city for about 8km (Figure 1[7]). It shows a systematic use of bricks and suggests a large-scale production of it. If the exterior walls represent the prevalence of stone, the heart

---

[1] CReA-Patrimoine/Research Centre for Archaeology and Heritage, Université libre de Bruxelles.
[2] We would like to thank Dr Michel de Waha (CReA-Patrimoine/ULB) and Dr Paulo Charruadas (CReA-Patrimoine/ULB) for all their valuable advice and our fruitful discussions about building materials. We are also grateful to Catherine Ostrow and Britt Claes for all the help with the translation of this article.
[3] See especially Debonne 2010, 11–34; Emmens 2008, 73–114; van Tussenbroek 2008, 115–134.
[4] Mentioned in Tijs 1999, 77–78; For Brussels, the works of Guillaume Des Marez and Victor Gaston Martiny offer the best synthesis of this architecture: Des Marez 1928 and Martiny 1992.
[5] See in particular the work led by Krista De Jonge and Koen Ottenheim on the architecture of the Southern and Northern Netherlands: De Jonge and Ottenheim (eds) 2007.
[6] This research is conducted within the scope of a doctoral thesis on the archaeological study of building materials in Brussels, led by Dr Michel de Waha: Sosnowska 2013. For the study of architectural ceramics, this doctoral thesis will propose and provide a new lecture of the brick phenomenon in Brussels with three mains parts. The first one proposes an approach on brick kilns in Brussels. The second one proposes the material study of bricks: dimensions, traces of moulding and archeometrical determination of the composition to understand the evolution of the fabrication recipe and the importation of foreigner products. In this case, an important collaborative project was created with Dr Eric Goemaere, Dr Thierry Leduc (Geological Survey of Belgium - Royal Belgian Institute of Natural Sciences) and Dr Mark Golitko (Field Museum, Chicago) for the archeometrical study of the material. The third chapter, finally, explores the implementation of bricks in brussels architecture.
[7] MRBC-DMS: Ministère de la Région de Bruxelles-Capitale – Direction des Monuments et des Sites (Brussels Capital Region – Heritage Direction); MRAH: Musées royaux d'Art et d'Histoire (Royal Museums of Art and History).

Figure 1. Porte de Hal, last remain of the second city wall defending Brussels and one of the seven city gates, dated from the second half of the 14th century (Modrie S. © DMS/MRBC)

Figure 2. Porte de Hal, details of the barbican defending the gates (Sosnowska P. © CReA-Patrimoine/ULB)

of the masonry is essentially made of brick (Figure 2). We can cautiously propose that this practice is the result of the lower quality of brick over stone (resulting from the recipe and / or firing) and its high porosity that makes it not frost resistant and therefore a more fragile material[8]. According to the iconography, we can assume that the whole fortification was built using the same implementation. We can note here that the first outer wall of Brussels was entirely built of stone and earth, a century earlier.

In the case of Brabant and Hainaut, historiography tends to closely associate the onset of massive brick production during the 14th century with the construction of major public works, especially urban townwalls[9]. J. A. Dupont has linked the existence of the communal brick kiln in Mons and its operating authority with the construction of the ramparts. The transition from an exploitation concession to private entrepreneurs (which explains the disappearance of data in the accounts of the city) would coincide with the end of a major construction phase of the townwall. The argument put forward is that the use of this new material appears to be an expedient for reducing construction costs[10]. It could also be argued that the time and quantity ratio of the production played a role in the choice of its implementation.

If we follow this assumption, it seems necessary to qualify it in the light of the archaeological research. Considering the available literature, the use of brick in the city of Brussels could be prior to the construction of the second townwall. In fact, archaeological excavations have revealed some brick structures built before the second half of the 14th century. It is hard to imagine that the city would engage in the construction of a townwall with an «experimental» material, without knowing its qualities. The protection of the interior brick wall with a stone facing at the Porte de Hal tends to suggest that the porous properties of bricks were well known and that masons knew how to counter this disadvantage. In this case, the designers of the townwall had some experience with the material.

At the convent of the Riches-Claires, two occupation phases exposed brick masonry structures. The first one is part of a wall and two brick floors combined with tiles. The set dates between the second half of the 13th century and the first half of the fourteenth century[11]. The second one is of a quite different scale, since it is the first «real» house built in hard materials[12]. This set includes floor and walls, properly built in brick, and is dated between 1306–1310 and 1356[13]. The excavation of the rue de Dinant also revealed some brick remains, including postholes filled by a blocking of broken bricks, dated in the late 13th or early 14th century[14] and a series of walls dated in the first half of the 14th century[15]. In the latter case, the preserved elements are of a minimal extent.

These few cases clearly show the existence of brick

---

[8] Piers 2005, 48–49, 59. Paris, Eyrolles.
[9] de Waha 1986, 52–59.
[10] de Waha 1986, 54.
[11] De Poorter 1995, 146–149.
[12] De Poorter 1995, 63–69, 150–151.
[13] De Poorter 1995, 150–151.
[14] De Poorter 2001, 186, 219.
[15] De Poorter 2001, 188, 219.

Figure 3. East sidewall at 180 rue de Flandre, dated from the end of 14th or the beginning of the 15th century (Byl S., Charruadas P., Devillers C. Sosnowska P. © DMS/MRBC – CReA-Patrimoine/ULB)

Figure 4. Details of the bricks used in the sidewall at 180 rue de Flandre (Byl S., Charruadas P., Devillers C. Sosnowska P.. © DMS/MRBC – CReA-Patrimoine/ULB)

production prior to the second townwall. If its magnitude is still difficult to determine, we should nevertheless highlight that these buildings are not linked to a particularly prominent social class, nor to modest housing. In this case, we need to highlight the importance of wooden architecture. All the findings suggest the coexistence of these two architectural types: one of wood and one of brick. A ratio between the two types – only based on the archaeological published data – remains unpredictable for the moment. The main argument is, nevertheless, that the wooden structures, usually identified by negative post hole traces as well as the incomplete aspect of plans of the raised structures do not allow for a clear identification of the function of all encountered structures.

Several 14th century brick buildings are still partially preserved. The priory church of the Rouge Cloître in Auderghem (1381), for exemple, shows interior brick sidings. In this case, the work is contemporary with the townwalls[16]. We should add the recent discovery of the sidewalls of a house located along the former steenweg (currently rue de Flandre), of which the inner facing is entirely in brick (Figure 3), while the outer facing involves sandstone and brick[17]. This building is dated from the 14th century or the first decade of the 15th century[18]. Furthermore, a barrel-vaulted cellar was partially unveiled during the excavations conducted on the square of the Vieille-Halle-aux-Blés and the rue du Chêne. It concerns two cellars that could date back to the second half of the 14th or the early 15th century[19], which corresponds to the period of urbanization of the neighborhood[20]. In a house situated 4 Place du Grand Sablon, brick sidewalls with identical formats to those of the Porte de Hal have been discovered and could therefore go back to the 14th

century[21]. We should note that the first mention of brick factories is dated from the late 14th century and informs us of intra muros locations, certainly on very sparsely built land[22].

Through these examples, it appears that the manufacture and use of brick does not go back beyond the 14th century, except perhaps in the cases where the wide range goes down to the second half of the 13th century. How to conceive the size of this production? Should we see a yield and a two-speed development, one prior to the construction of the townwall, the other resulting from the execution of this huge project? Presumably, it was a driving force in the adaptation of the architecture to the «new» material. Was it also a factor of innovation in the quality of the brick and the means of production? It is difficult to say for the moment, because no samples of brick were removed during excavations on the mentioned sites, with exception of the rue de Flandre and the Porte de Hal.

For these two buildings, the difference in brick quality is quite surprising (Figure 4 and see Figure 2). In the case of the Porte de Hal, the visible interior facing bricks are regular, well fired, whole for the majority. It does not mean the same for the rue de Flandre, where they are essentially incomplete. It is, in our view, less a difference in quality of the production than in the choice of the supply: for the house of the rue de Flandre, second choice and probably cheaper materials where used. Therefore, it is not without interest to see them used in a house that is not distinguished by its exceptional wealth, or by special poverty. Eithermore, it does not appear that we should detect a chronological clue, given the quality of the masonry observed at the

---

[16] Wauters 1855, 355–356; Modrie 2010, 89.
[17] Byl, Charruadas, Devillers and Sosnowska 2011.
[18] Byl et al. 2011, 37–38.
[19] Diekmann 1997, 109–114; Braeken and Van Eenhooge 1991.
[20] Billen and Thomas 1997, 105–108.
[21] de Waha, s.d.
[22] Moreover, this situation will remain unchanged until the 18th century. One must indeed expect to see in 1776 the first ordonance prohibiting the construction of brick factories in the city: A.V.B. Correctie boeck, no1094, t.10, fo104, May 10, 1776, quoted by Victor Gaston Martiny in: Martiny 1992, 20.

Riches-Claires, although anterior by little more than half a century and dated between 1306–1310 and 1356[23].

Some remarks on the use of tiles in Brussels before the 14th century

As for bricks, the history of Brussels tiles – and even floor tiles – remains to be written. We will therefore limit our approach to the use of tiles before the 14th century.

For the given chronological context, archaeological excavations have revealed two main types of tiles. The first one contains the tegula and imbrex type. Excavations on the rue de Dinant have revealed this type of tiles in a context dated to the first half of the 13th century[24]. A fragment of a Roman style tile was picked up at the top of a filling of a pit dated to the second half of the 13th century[25]. Finally, the excavations of the church of Saint-Pierre de Neder-Over-Heembeek by J. Mertens during the month of March 1959 in the context of the Romanesque church[26] revealed a number of tile fragments of the tegula and imbrex type.

The second typology, the most important one, is defined by thin flat tiles, of rectangular shape, provided with a hook and / or a hole to attach to the roof boarding. The oldest examples clearly date back to the 13th and 14th centuries. At the Riches-Claires, these tiles were installed in a pavement floor, also made of brick[27]. The structure is dated between the second half of the 13th century and the first half of the 14th century[28]. Tiles are still present in two excavated pits on the rue du Vieux-Marché-aux-Grains, which filling includes nails, stone, mortar, ceramics, etc.[29]. In one pit, the tiles were glazed. Overall, the material of the first pit can be dated between the second half of the 13th century and the mid-14th century[30]. In the rue de Dinant, a layer of fill showed a concentration of glazed and unglazed tiles, dating from the first half of the 14th century[31].

Other findings highlight the existence of roofing materials in architectural ceramics. In some cases, it is difficult to determine whether it concerns the tegula and imbrex type or flat tiles. This is for exemple the case of the impasse du Papier, where fragments of washers or washers cut into sometimes glazed tiles were identified in a historical context, dated between the second half of the 13th and the middle of 14th century[32]. In the rue du Vieux-Marché-aux-Grains, a piece of tile was found in a context of the late 12th century[33]. We should also note the discovery of a fragment of a ridge tile in orange clay with a brown glaze and the

Figure 5. Coudenberg Palace, view of the facing stone of the Aula Magna constructed between 1452 and 1460 (Sosnowska P. © CReA-Patrimoine/ULB)

beginning of a decorative motif found in the impasse du Papier[34]. It has been dated to the late 13th or early 14th century[35].

The presence of this tegula and imbrex type resolutely leads to questions. The contexts in which they were found are all post-Roman. However, it does not seem that these objects originate from re-used roman structures. We can therefore assume a medieval allocation of these fragments, prior to the 14th century. During this same period flat tiles seem to appear. The discoveries made in Flanders confirm this diagnosis. Here, their use is attested from the 10th to the early 13th century, during which time they were replaced by flat rectangular format tiles[36]. The same observations were made in England, where Roman type tiles were found in the 12th and 13th centuries, before the production changed during the 14th century[37]. It is interesting to note that the introduction of these rectangular tiles is concomitant with the first remains of brick buildings. This finding requires a

---

[23] De Poorter 1995, 150–151.
[24] De Poorter 2001, 182, 219.
[25] Massart 2001, 285, 298.
[26] Van Bellingen 2011, 83.
[27] De Poorter 1995, 150–151.
[28] De Poorter 1995, 150–151.
[29] Siebrand, Demeter and De Poorter 2001, 139–143.
[30] Siebrand 2001, 142–143.
[31] De Poorter 2001, 188, 219. This phase also dates postholes whose filling or blocking is dated to the first half of the fourteenth century: De Poorter 2001, 188, 219.
[32] Massart 2001, 283–285 and 298.
[33] Siebrand et al. 2001, 132.

[34] Massart 2001, 286–288.
[35] Massart 2001, 298.
[36] Debonne 2010, 14. The author cites the sites of Petegem, of Ename, of Dikkelvenne, of Alost and Moorsel.
[37] Drury 2000, 58.

Figure 6. 15 Grand-Place, view of the cellars combining strong transverse stone arches supporting a brick vault (Sosnowska P. © CReA-Patrimoine/ULB)

further reflection on the appearance of these two products, relatively similar in their production. This reflection will be subject to further research.

The use of brick from the 15th century to the 18th century

The use of brick in the 15th century

Brussels' architectural heritage retains some visible and easily accessible remains from the 15th century architecture. Monumental buildings, churches, palaces and public buildings constitute the bulk of the evidence from this period. However, their appearance badly reflects the prominence of brick in the art of the construction during this golden century for the city. Stone ornaments generally hide the heart of the brick masonry, for interior domestic facings as well as for the division walls. For the city of Brussels, the Aula Magna (Figure 5), the current Cathedral of Saints-Michel-et-Gudule, the churches of Notre-Dame du Sablon and Notre-Dame de la Chapelle and the Town Hall are the best known examples. In all of these cases, brick was used for the arches, the builders taking advantage of the lightness and the strength of the material. It should

be noted that in the case of the Aula Magna, the use of brick was granted by Philippe le Bon in order to reduce the cost of its construction[38]. In return, the city, which funded the work, had to commit to whitewash the brick walls[39]. In a more rural area, the Carthusin monastery of Scheut in Anderlecht[40] may have been built on this model, with outside walls made of stones and inside walls built with bricks.

The so-called houses of the Colline and the Pot d'Etain situated Grand-Place and built during a monumental project implemented by the City of Brussels in 1440 (current Duke of Brabant), include these features. The 1695 bombing by the French armies has profoundly changed the appearance

---

[38] Dikstein-Bernard 2007, 43–44; it should be noted that at the current Palais Rihour in Lille, the city had imposed this material for which the technique was unknown: De Jonge 2003, 93–134. One should also note that before the major construction phase, various constructions were carried out at the Palace of the Duke of Brabant. Thus, Henne and Wauters reported that in 1403, 8000 tiles were bought to pave the great hall, the chapel, the rooms of the Duchess, those of lords Rotselaer, of Seyne, of Immerseel, the furriers, the clerics or secretary room, the recipients room (...) and in 1415, 8000 bricks were bought to pave the smaller rooms: Henne and Wauters 1845, 319–320.
[39] Dikstein-Bernard 2007, 44.
[40] de Waha 1979, 264.

of the Grand-Place, but the archives of the corporation of carpenters still attest of its former state[41]. Today, only the cellars still reflect the duality combining strong transverse stone arches supporting a vault made of bricks (Figure 6). It is possible that other types of coverage existed, but we only have a very incomplete concrete picture of this type of structure. This consideration also applies to the 16th century.

These examples should not obscure the existence of some monuments that show the overall use of this material. This is for exemple the case for the hotel Clèves-Ravenstein, whose 19th century restorations made by Saintenoy did not erase both hurdles of white stone sustained by the brick facade[42]. Similarly, the prestigious hotel Lalaing was built from houses mixing brick and stone in their facades. This architectonic structure of the facade consists of plinths in white stone (less moisture sensitive), which support a brick wall. The latter was reinforced by stones at height of the corners and punctuated by bands of white stone. Furthermore, bays are often drilled with stone framings[43]. This typology leaves an indelible mark on the architecture of our region until the 18th century. Isn't this to be considered traditional architecture par excellence? This state is still reflected in the representation of the Church of Sainte-Gudule dated from the late 15th century. The painter distinguished the materials of the buildings well and the representation that he gives of them perfectly fits with the above-mentioned scheme[44].

Archaeological excavations also deliver numerous structures dating from this period. The remains are nevertheless rare according to the various available and / or published excavation reports. The excavations of the rue d'Une Personne revealed the existence of a building dated in the early 15th century, reflecting the late urbanization of the ground. The masonry combines stone and brick, but its appearance is sloppy[45]. The most impressive discoveries are all three houses unearthed along the rue Vander Elst and adjacent to the former convent of the Pauvres-Claires[46]. The street facades of these homes could not be reached during the investigations. The rest of the walls (facades and partition walls) were made of brick. At 180 rue de Flandre, a sidewall of the second quarter of the 15th century could have been revealed, if one follows the 14C dates obtained on a sample of charcoal taken from the mortar[47]. The frontwall was characterized by a white stone plinth from which departed a brick wall, alternated by brick courses and stone bands. Finally, the excavations of the hotel de Merode revealed a first core of stone and brick built prior to 1515–1516[48].

In some cases, the range for dating remains loose due to the lack of chronological markers. This is the case of the Dewez house in which a brick core, dating from the 14th or 15th century, was unearthed[49]. The same difficulties were encountered on the site of the postal relay of the Vieille-Halle aux Blés[50]. In addition, the unearthed structures often do not specify whether they were entirely made of brick or if they met the typology of houses with "petrified" gutter walls or the one of the so-called three quarters houses. It is also from this period during which Charles Buls defined the stylistic evolution of gables in Brussels[51]. The latter was partially corrected by Victor-Gaston Martiny, with the addition in the 15th century of a stepped gable and cord «in Antwerp style,» observed on a house now demolished rue Bodenbroeck[52].

In rural architecture, the references of brick are rare. The only case remaining is the one of the chimney in the Béguinage of Anderlecht (Brussels), which could date back to the same construction phase of the Béguinage in the first half of the 15th century[53]. In Saints-Pierre-et-Guidon in Anderlecht, brick gutter walls are apparent on the sidewalls of the nave and in the vaulting hidden under the coatings, building phase dated back to the 15th century[54]. The same goes for the church of Saint-Denis in Forest (Brussels), including the gable between the nave and the choir and the lower parts of the west tower[55].

Finally, the frequent use of brick is highlighted in the written documentation of the 15th century. An ordinance of July 1422 regulates the manufacture, while others, in 1443, then in 1486, regulate the price[56]. The ordinance for the surveyors of 1451 does not explicitly mention the use of this material. Only the Article 47 refers to the material by the term steen that could refer to stone as well as brick[57]. The construction accounts of the east wing of the townhall of 1405 mention the persons who worked on its construction, whether suppliers or manufacturers as well as the materials for which they were responsible. We learn that there are distinct brick products: quareelen or bricks, bricks used for chimneys or hertquareelen or klinkaert[58].

The use of brick in the sixteenth century

Material evidence of brick architecture in Brussels is

---

[41] Dickstein-Bernard 2008, 9.
[42] Martiny 1992, 20.
[43] De Jonge 2007, 56.
[44] Maître de la vue de Sainte Gudule, L'instruction pastorale ou la prédication de Saint-Géry, end of 15th c., Paris, Musée du Louvre, inv. 1991.
[45] Diekmann 1997, 25.
[46] Claes 2006, 29–31.
[47] Byl et al. 2011, 46.
[48] Sosnowska 2007, 89.
[49] Sosnowska 2011, 172–173.
[50] Diekmann 1997, 117.
[51] Buls 1908.
[52] Martiny 1991, 115.
[53] Charruadas and Sosnowska (2013a, 26-28). The Erasmus House in Anderlecht (Brussels) could perhaps keep a building core prior to 1515. An archaeological investigation is expected to support this hypothesis. Martiny also reported the construction of a bread oven in the fifteenth century in the farm of the Hof ter Biest (Brussels): Martiny 1973, 191; voir également de Waha 1979, 264.
[54] Communication of Mr Michel de Waha.
[55] Communication of Mr Michel de Waha.
[56] Martiny 1992, 20. The author cites the Perkement boeck preserved in the Archives of the City of Brussels for three ordinances: AVB, IX, f°10 for the ordinance of 1422, idem, 83, f°2 for the one of 1443 and, idem, 106, f°8 for the one of 1486.
[57] de Waha 2000, 71.
[58] Bernard 1959, 272.

somewhat more available for the 16th century. Recent research has led to a series of discoveries of prime importance to understand the evolution of the 16th-century habitat, its material and implementation, as well as the type of the so-called « bourgeois homes », to say it in the words of Victor-Gaston Martiny[59]. The study of urban palaces, like the ones of the Mansfeld (current hotel de Merode) and the Lalaing also provide valuable insights. However, we must keep in mind that at that time, as in the 15th century, wooden architecture is still quite present in the urban landscape.

The remains of some urban palaces suggest that 16th century aristocratic architecture is basically made of brick, as evidenced by the hotel of Hoogstraeten and some wings of the current Hotel de Merode[60]. It concerns a few cores of the 16th century without precise chronological anchor, except for a first core dating between 1515 and 1516d[61]. The use of white stone, although still important, is limited to the structural elements of the building. In the facade, it is implemented for the plinth, window frames and doors – single or monumental – the dripstones and putlog holes (Figure 7). Inside, it is carved in chimney cheeks and mantels, beam corbels or stair-steps.

In contrast, the construction of the chapel of the imperial palace follows the previous construction method: a stone facade and an interior brick facing. The mighty octagonal pillars supporting the arches have a stone facing and a heart filled with fragmentary brick and rubble, all poured into the mortar. Again, the interior walls are built mainly with bricks, however, they are placed on a series of foundation stones. Structuring elements, harpages bays or transverse arches supporting brick arches are, in turn, of white stone.

As for the "bourgeois" house of the 16th century, it remains comparable to prior periods as an architecture mixing wooden facades and brick and stone gutter walls and rear facades. If the timber frames no longer appear on the front, their existence shines at the truss and therefore at the walls that carry them. Examples of the 120 rue de Laeken, 180 rue de Flandre, 4 and 132 rue Haute are explicit about it[62]. The presence of brick facades going back to those times is sometimes still visible throughout the year anchored in the walls. Thus, 26 rue Sainte-Catherine could go back to 1597. In this context, it should be noted that ordinances prohibiting the construction of wooden facades and even the maintenance of them, date respectively[63] from 1566 and 1567.

We should note that from the 14th century to the 16th century, the implemented formats are 26–29x12–13, 5x5–7cm. The decrease in formats as it has occurred in parts of present-day Flanders and the Low Countries could not be observed for Brussels. As we will see, this caliber

Figure 7. Merode Palace, view of the facade of the southeast aisle dated from the 16th century (Sosnowska P. © MRAH-DMS/MRBC)

Figure 8. Types of bricks used in Brussels between the 15th and the 17th century. The first (c. 16th) and the second one (c. 17th) are produced in Brussels. The last one, the upper one, dated between 1789 and 1799, is maybe an import product (Sosnowska P. © CReA-Patrimoine/ULB)

---

[59] Martiny 1991, 109.
[60] Demeter and Sosnowska 2007, 49–54.
[61] Demeter and Sosnowska 2007, 49–54; Eeckhout 2006, 4.
[62] Charruadas and Sosnowska, 2013b, 3–13
[63] Martiny 1991, 110.

Figure 9. 120 rue de Laeken, view of the Flemish bond dated between the 15th and the 16th centuries (Sosnowska P. © CReA-Patrimoine/ULB)

will remain quite stable until the second half of the 18th century (Figure 8). Our conducted research demonstrates the presence of a particular brick bond for the 16th – and probably 15th – century, which completely disappears in the following century. It concerns the Flemish bond[64], that could be observed at 73–75 and 120 rue de Laeken (Figure 9), 132 rue Haute, 19 rue Notre-Dame du Sommeil, the Erasmus house, 59 and 180 rue de Flandre, rue Villa Hermosa, 19 rue du Marché-aux-Fromages, as well as at 2 and 4 rue de Flandre. The implementation is, to this day, only identified on walls with a thickness of one brick (c. 27,5cm). The use of this bond could be specifically related to the change in brick format produced in Brussels and whose recorded variability sometimes reaches 3 centimeters for the same building phase.

Finally, the first mention of an import authorization for lime and brick appears in 1551[65]. This order is dated ten years before the digging of the canal linking Brussels to the river Rupel, giving access to the rich brick kiln region of Boom. This opening has certainly facilitated the importation of materials, like other products elsewhere. Currently, we unfortunately find it difficult to identify material for the given period of these imported products. No brick has been so far identified as being imported, which is not the case for the later periods.

The use of brick in the seventeenth and eighteenth centuries

The current Brussels built landscape is deeply influenced by the 17th century. In view of the entire architectural documentation, it may be regarded as the century of full brick use. The remains and testimonies are numerous and cover the whole of the architectural production: religious buildings, aristocratic, bourgeois and smaller habitats. One can also assume, given the numerous preserved gables stylistically dated to this period, that the 'petrification' of the facades had to be intense. Such changes were observed at 4 and 180 rue de Flandre and at 4 and 132 rue Haute. Surveys in some areas of the city, including the rue Haute, seem to attest of the existence of other examples (Figure 10). However, we must remain careful because no archaeological investigations have been conducted in the field.

For this period, we have observed only little changes in the implementation of materials, with the exception of the brick bond which appears always implemented in its crossbond form[66]. The builders continue the execution of the structural work in a traditional way, bricks constitute the main material and the use of stone is concentrated on specific elements such as door frames, window frames and plinths. In contrast, the format of the brick shrinks somewhat. The caliber is essentially between 26/27x12/13x5/5,

---

[64] Bond consisting of an alternating stretchers and headers repeated continuously on each course.
[65] Het Gheel correctie boeck, 196, 354 fo vo, ordinance of May 29, 1551 as cited in Martiny 1992, 20 and note 42.

[66] A masonry bond in which a course of stretcher bricks alternates with a course of header bricks. The joints of one stretcher course come midway between those of the stretcher courses above and below.

Figure 10. Facade at 180 rue de Flandre, dated from the 17th century but rebuild against a core habitat from the end of 14th or the beginning of the 15th century (Byl S., et al. © DMS/MRBC – CReA-Patrimoine/ULB)

Figure 11. Sidewall at 50 rue Marché-aux-Herbes wich is composed by bands of local products, re-used material or new, and bands of import products (Modrie S., Sosnowska P. © MRBC/DMS - CReA-Patrimoine/ULB)

5cm. Variability is lower and therefore reflects a more homogeneous production that must be linked to a change in the recipe for brick making.

Significant changes were observed during the reconstruction that followed the bombing of the city of Brussels in 1695. They include all materials. The demand for material was such that bricks, used to rebuild the affected buildings, were massively imported from other regions. The archives mention bricks from Boom, Holland and IJssel, in parallel with a local production[67]. The archaeological research conducted for this article helped to highlight these different products. The use of materials from different origins and with different formats had an impact on the structural work, leading sometimes to specific implementations (Figure 11).

In the 18th century, the use of imported bricks is evidenced by some construction archives and ordinances: imported bricks were subjected to control by an expert who judged their quality[68]. The major archaeological evidence, however, doesn't date back before the second half of this century. Until, only one case shows the use of imported bricks in the rue des Pierres 34. Bricks of local production and imported bricks are used at the same time and even next to each other in the same masonries. The use of brick still appears as the main element in construction in comparison with stone, but this should not obscure the fact that a fully parallel architecture of stone was equally present.

Conclusions

Brick is one of the major building materials in Brussels. In general, its use seems to exceed by far that of stone, which remains essentially attached to monumental architecture of the 14th, 15th and 16th centuries, as well as to structural elements of the facades and equipments.

---

[67] Culot, Hennaut, Demanet and Mierop 1992, 264–265; Hennaut 2011, 76–78.

[68] AVB, Recueil de proclamations de Kuyghe, t. 16, p. 392, Ordinance of 1 March 1759 confirmed 13 September 1797, cited in Martiny 1992, 20 and note 40. AVB Correctie boek, No. 1094, t.10, fo104, Ordinance of 10 May 1776, quoted in Martiny 1992, 20 and note 40.

Its appearance is later than in the county of Flanders and in the northern Low Countries, as the first archaeological evidence doesn't go back beyond the 14th century, except perhaps one case of the second half of the 13th century. It is therefore necessary to link the idea of brick appearance in Brussels with the establishment of large urban projects, particularly the construction of the second townwall. These testimonies assume a local production, but today we cannot yet define the extent and quality, nor the status of the contractors who were responsible for it. We can cautiously say that it concerns a relatively upper social group. Brick production requires indeed specialized trades for its manufacturing, infrastructure, tools and fuel. It also appears that the general shape of the tiles follows the one of the brick and that their emergence seem concurrent.

The construction of the second townwall certainly played a role in the development of brick in Brussels. The size of the project has indeed requested a large supply in all types of material. This project has enabled a practical implementation on a large scale, perhaps in contrast to other sites where the use of brick was on a smaller scale, such as the Town Hall and the Aula Magna. Other cases show instead a use of it in the frontages, as evidenced by the hotel Clèves-Ravenstein. For this site, brick is associated with stone, located on the structural parts of the facade. It should be noted that it is this type of implementation that lasted until the eighteenth century, with, of course, variations reflecting the taste of each period. It certainly reflects traditionalism in the Brussels constructions. The use of brick appears increasingly important during the 16th century. It will be fully implemented by the aristocracy. For the bourgeois home, we saw that the wooden architecture retained an important place, although the gutter walls were more and more «petrified» and rebuilt with stronger material. It seems that during this century the recipe of bricks changed but not the format which was to decrease only in the 17th century. This last one seems to be the century of the systematic use of brick. New buildings like the cloister of the convent of the Riches-Claires for example show a significant use of the material. Facades are built or rebuilt in brick as well as all the structures linked to fire. Finally, in the 17th century, perhaps even in the second half of the 16th century, we see the emergence of imported bricks. Their identification remains to this day largely confined to the reconstruction of Brussels following the bombardment in 1695. Their use seems important, as there are many examples.

In the 18th century, these imports seem at first less important, with only one example being reported for the first half of this century. This decrease is most likely related to the specific conditions experienced in the aftermath of the siege of the city in 1695, where supply material requirements had to be significant. The magnitude of import is more important during the second half of the 18th century, during which local bricks and imported bricks appear together in the same masonry. It is probably in the last quarter of this century, or even in the beginning of the next century, that one sees a final disappearance of the traditional Brussels format in favor of a smaller brick matching those produced particularly in the north part of the country.

References

Billen, C. and Thomas, F. 1997. Enquête historique sur le quartier de la place de la Vieille-Halle-aux-Blés. In Diekmann, A., Artisanat médiéval et habitat urbain, rue d'Une Personne et place de la Vieille-Halle-aux-Blés. Archéologie à Bruxelles 3, 105–108. Brussels, Ministère de la Région de Bruxelles-Capitale. Direction des Monuments et des Sites.

Braeken, J. and Van Eenhooge, D. 1991. La Maison Schott. Brussels. Bruxelles, Fondation Roi Baudouin.

Buls, C. 1908. Le Vieux-Bruxelles. Travaux élaborés par le Comité institué sous le patronage de la Ville de Bruxelles et de la Société d'Archéologie de Bruxelles, vol. 1: Préface-programme; vol. 2: évolution esthétique. L'évolution du pignon à Bruxelles. Bruxelles, Van Oest et Cie.

Byl, S., Charruadas, P., Devillers, C. and Sosnowska, P. 2011. Étude archéologique du bâti d'une habitation sise rue de Flandre 180 à 1000 Bruxelles [BR247]. Évolution d'une habitation du Moyen Âge à nos jours (14e – 21e siècle). Unpublished report, Université libre de Bruxelles and Direction des Monuments et des Sites du Ministère de la Région de Bruxelles-Capitale.

Charruadas, P. and Sosnowska, P. 2013a. Petit béguinage et architecture vernaculaire: Étude archéologique d'un pan-de-bois du XVe siècle conservé dans l'actuel musée du béguinage d'Anderlecht. Revue belge d'archéologie et d'histoire de l'art 20, 5–44.

Charruadas, P. and Sosnowska, P. 2013b. 'Petrification' of the architecture in Brussels: An attempt at explanation: between Construction Methods, Building Materials and Social Changes (c. 13th–17th). In 17th Vienna Conference on Cultural Heritage and New Technologies: Urban archeology and Excavations: Urban archeology and excavations to Reach and to Unveil the Hidden Spirit of the Town (5–7 November 2012: Vienna). Vienna, Museen der Stadt Wien Stadtarchäologie. Available at: http://www.stadtarchaeologie.at/?page_id=5326

Claes, B. 2006. Archeologisch onderzoek van het voormalige Arme Klaren klooster te Brussel», Archaeologia Mediaevalis. Chronique 29, 29–31.

Culot, M., Hennaut, E., Demanet, M. and Mierop, C. 1992. Le bombardement de Bruxelles par Louis XIV et la reconstruction qui s'en suivit 1695–1700, Bruxelles, Archives d'Architecture Moderne.

Debonne, V. 2010. Bouwen met baksteen in het graafschap Vlaanderen, ca 1220–1440. Een overzicht. In Oost, T. and Van de Voorde, E. (eds), Vuur en Vlam. Omgaan met baksteenerfgoed in Vlaanderen, 11–34. Antwerp, Provincie Antwerpen.

De Jonge, K. 2003. Bourgondische residenties in het graafschap Vlaanderen. Rijsel, Brugge en Gent ten tijde van Filips de Goede. Handelingen der Maatschappij voor geschiedenis in oudheidkunde te Gent 54, 93–134.

De Jonge, K. and Ottenheim, K. (eds) 2007. Unity and discontinuity: architectural relationships betweens the Southern and Northern Low Countries from 1530 to 1700. Architectura Moderna 5. Turnhout, Brepols.

De Jonge, K. 2007. Antiquity Assimilated: Court Architecture 1530–1560. In De Jonge K. and Ottenheim, K. (eds), Unity and discontinuity: architectural relations between the Southern and Northern Low Countries 1530–1700. Architectura Moderna 5, 55–78. Turnhout, Brepols.

Demeter, S. and Sosnowska, P. 2007. Sur les traces des comtes de Mansfeld à Bruxelles, les vestiges archéologiques découverts dans l'hôtel de Merode. In Mousset, J.-L. and De Jonge, K. (eds), Un prince de la Renaissance. Pierre-Ernest de Mansfeld (1517–1604). II Essais et catalogue, 49–54. Luxembourg, Musée National d'Histoire et d'Art.

De Poorter, A. 1995. Au quartier des Riches-Claires: de la Prieumspoort au couvent. Archéologie à Bruxelles 1. Brussels, Ministère de la Région de Bruxelles-Capitale. Direction des Monuments et des Sites.

De Poorter, A. 2001. Het archeologisch onderzoek op een terrein in de Dinantstraat (1995). In Blanquart, P., Demeter, S., De Poorter, A., Massart, C., Modrie, S., Nachtergael, I., Siebrand, M. (eds), Autour de la première enceinte, Archeologie à Bruxelles 4, 178–225. Brussels, Ministère de la Région de Bruxelles-Capitale. Direction des Monuments et des Sites.

Des Marez, G. 1928. Guide illustré de Bruxelles en deux volumes I. Les monuments civils. Bruxelles. Brussels, éditions du Touring Club de Belgique.

de Waha, M. 1986. Aux origines de l'architecture de briques en Hainaut. In Deroeux D. (ed.), Terres cuites architecturales au Moyen Age, actes du Colloque des 7–9 juin 1985, Musée de Saint-Omer, 52–59. Arras, Commission départementale d'histoire et d'archéologie du Pas-de-Calais.

de Waha, M. 2000. L'ordonnance de 1451 et le paysage urbain bruxellois. Première esquisse. In Bulletin de la Commission Royale pour la publication des anciennes lois et ordonnances de Belgique, XLI, 59–78.

de Waha, M. (without dates). Grand Sablon 4. Rapport commandé par le propriétaire. Unpublished report Université libre de Bruxelles and Institut de Gestion de l'Environnement et de l'Aménagement du Territoire.

Diekmann, A. 1997. Artisanat médiéval et habitat urbain, rue d'Une Personne et place de la Vieille-Halle-aux-Blés. Archéologie à Bruxelles 3. Brussels, Ministère de la Région de Bruxelles-Capitale. Direction des Monuments et des Sites.

Dikstein-Bernard, C. 2007. La construction de l'Aula Magna au palais du Coudenberg. Histoire du chantier (1452–1461). Annales de la Société Royale d'Archéologie de Bruxelles 68, 43–44.

Dickstein-Bernard, C. 2008. La maison édifiée en 1441 sur la Grand-Place de Bruxelles par le métier des charpentiers, élément d'un ensemble architectural de six maisons en pierre conçu par la ville. Revue belge d'archéologie et d'histoire de l'art 77, 3–26.

Drury, P. 2000. Aspects of Production, Evolution and Use of Ceramic Building Materials in the Middle Ages. Medieval Ceramics 24, 56–62.

Eeckhout, J. 2006. Rapport d'analyse dendrochronologique. L'hôtel de Merode à Bruxelles. Analyse complémentaire 4. Unpublished report University of Liège and Ministère de la Région de Bruxelles-Capitale. Direction des Monuments et des Sites.

Emmens, K. 2008. De oudste Friese baksteen. Een heroriëntatie op de introductie en vroege toepassing van baksteen in Friesland en Groningen. In Coomans, T. and van Royen, H. (eds), Medieval Brick Architecture in Flanders and Nothern Europe: the question of the Cistercian origin. Novi Monasterii 7, 73–114. Koksijde, Ten Duinen

Hennaut, E. 2011. Le chantier de la reconstruction: organisation, matériaux et techniques. In Heymans, V. (ed.), Les maisons de la Grand-Place de Bruxelles, 76–78. Bruxelles, CFC édition.

Henne, A. and Wauters, A. 1845. Histoire de la Ville de Bruxelles 3, Bruxelles, Librairie Encyclopédique de Perichon.

Martiny, V.-G. 1973. Le concours annuel de relevés de la Société Centrale d'Architecture de Belgique. Folklore Brabançon 198, 131–192.

Martiny, V.-G. 1991. La maison bourgeoise unifamiliale à façade étroite du 16e à l'aube du 20e siècle à Bruxelles. In Baetens, R. and Blonde, B. (eds), Nouvelles approches concernant la culture de l'habitat. New Approaches to Living Patterns. Actes du Colloque international de l'Université d'Anvers (24–25 oct. 1989), 109–146. Turnhout, Brepols.

Martiny, V. G. 1992. Bruxelles. Architecture civile et militaire avant 1900. Braine-l'Alleud, Collet J.-M.

Massart, C. 2001. Étude archéologique de l'impasse du Papier (1996). Blanquart, P., Demeter, S., De Poorter, A., Massart, C., Modrie, S., Nachtergael, I., Siebrand, M. (eds), Autour de la première enceinte, Archeologie à Bruxelles 4, 261–326. Brussels, Ministère de la Région de Bruxelles-Capitale. Direction des Monuments et des Sites.

Modrie, S. 2010. Les recherches archéologiques sur le site du prieuré de Rouge-Cloître. In Meganck, M., Guillaume, A., Atlas du sous-sol archéologique de la région de Bruxelles, Auderghem 21, 86–93. Brussels, Ministère de la Région de Bruxelles-Capitale. Direction des Monuments et des Sites.

Piers, G. 2005. La brique. Fabrication et traditions constructive. Paris, Eyrolles.

Siebrand, M., Demeter, S., De Poorter, A., 2001. Sondages sur le tracé du rempart, rue du Vieux-Marché-aux-Grains (1995). Blanquart, P., Demeter, S., De Poorter, A., Massart, C., Modrie, S., Nachtergael, I., Siebrand, M. (eds), Autour de la première enceinte, Archeologie à Bruxelles 4, 139–143. Brussels, Ministère de la Région de Bruxelles-Capitale. Direction des Monuments et des Sites.

Sosnowska, P. 2007. L'hôtel de Merode à Bruxelles: redécouvertes d'un site majeur (16e–17e siècles). In

Archaeologia Mediaevalis. Chronique 30, 87–91. Bruxelles.

Sosnowska, P. 2011. De la maison ordinaire à l'hôtel particulier. In Theuws, F., Tys, D., Verhaeghe, F. (eds), Medieval and Modern Matters, Archaeology and Material Culture in the Low Countries 2, 172–173. Turnhout, Brepols.

Sosnowska, P. 2013. De briques et de bois. Contribution à l'histoire de l'architecture à Bruxelles. Étude archéologique, technique et historique des matériaux de construction utilisés dans le bâti bruxellois (13e–18e siècle). Unpublished PhD, Université libre de Bruxelles.

Tijs, R. 1999. Architecture Renaissance et Baroque en Belgique. L'héritage de Vitruve et l'évolution de l'architecture dans les Pays-Bas méridionaux, de la Renaissance au Baroque. Bruxelles-Tielt, Racine.

Van Bellingen, S. 2011. Réexamen des fouilles archéologiques de J. Mertens sur le site de l'ancienne église Saint-Pierre à Neder-Over-Heembeek. In Meganck, M. and Guillaume, A. (eds), Atlas du sous-sol archéologique de la Région de Bruxelles. Neder-Over-Heembeek 23, 76–90. Brussels, Ministère de la Région de Bruxelles-Capitale. Direction des Monuments et des Sites.

van Tussenbroek, G. 2008. Vroege baksteen in Holland tot 1300. In Coomans, T., van Royen, H. (eds), Medieval Brick Architecture in Flanders and Nothern Europe : the question of the Cistercian origin, Novi Monasterii, 7, 115–134. Koksijde-Gent, Adbijmuseum Ten Duinen - Academia Press.

Wauters, A. 1855. Histoire des environs de Bruxelles ou description historique des localités qui formaient autrefois l'ammanie de cette ville 1. Bruxelles, Vanderauwera.

THE USE OF EARLY BRICKS IN SECULAR ARCHITECTURE OF THE 12TH AND EARLY 13TH CENTURY IN LÜBECK

Dirk Rieger

Abstract: During the largest excavation in the heart of Lübeck's historical centre new discoveries in the use of bricks in the secular architecture of the 12th and early 13th century were made. Bricks were normally used for erecting the sacral buildings and the city walls in these decades and seemed not to be common in secular constructions. Hitherto researchers assumed that there only had been two different phases of architecture in the mentioned time period: pure timber architecture of the 12th century that made way for the pure brick architecture of the 13th century. But now, for the first time, we could give evidence of the existence of a so-called "missing link" between these two construction types. Large wooden cellars were connected with stone and brick-lined elements like walls, stairways and wall-stringers. Some traces allow the interpretation of a fully storied first floor in brick masonry, which is absolutely unique and yet without any comparisons on the Baltic Sea.

Keywords: Archaeology, bricks, Germany, house building, Middle Ages, Lübeck, 13th century, timber cellar, 12th century.

Introduction

The excavation between the river Trave and the Church of St Mary in Lübeck (Figure 1) is one of the largest projects in a North German medieval town context, and the biggest excavation in the history of Lübeck's archaeological service (Figure 1). This area is defined by archaeological and historical research as the oldest quarter of Lübeck and is named the so-called "Gründungsviertel". The quarter had been one of the crucial points of the Hanseatic trade, with its densely built up plots, its centuries' old brick houses and cellars and its preserved medieval structures right in the heart of the historical centre between the harbour and the main market around the famous Lübeck Town Hall (Figure 2). In 1942 allied air raids destroyed about a third of the old town. This was considered as an act of revenge because the Nazis bombed Coventry two years earlier. After the war, the wasted land was reused as subsoil for modern school buildings in the 1950s and 1960s and as an inner-city car park. But with a plan for recreation in the centre of Lübeck, and for the preservation and protection of the UNESCO World Heritage, the pollution loaded buildings were broken down to erect a modern but again densely built up quarter. Therefore archaeological excavations had to be undertaken in four years on 44 medieval plots, all in all about 9000 square meters with a stratigraphy of two and a half to five meters. This enormous project could also be seen as one of the last chances to document the genesis of the earliest German Town on the Baltic Sea which later became the model for the foundation of other towns on the southern coastline and in the eastern parts of later Europe.[1] Financed by the German government, the UNESCO and the City of Lübeck, this project keeps up the archaeological research of the 1980s on neighbouring plots (Figure 1). Once this project has finished they will form one of the largest archaeological investigations of a medieval urban context in Northern Europe.

During the excavation archaeologists have recovered hundreds of extremely well preserved relics of wooden houses. These date to the founding years of the later Hanseatic League capital to the second half of the 12th century. During this period local Slavonic traditions were accompanied by new habits, brought by the Saxons, Frisians and Westfalians moving to Lübeck, which was founded in 1143. The newly discovered features and finds illustrate a duality of both local and inter-regional phenomena, particularly with the continuity and development of traditions.[2] But this article is neither about wooden relics nor pure timber architecture. It is about the first traces of using bricks as a new building material in secular architecture at the end of the 12th century to the first decades of the 13th century. This is period of economic growth and it is reflected in the laws and rights given to the Lübeck merchants by their town lords Henry the Lion Duke of Saxony and Bavaria, the Emperor Frederick I (known as Barbarossa) and later the Danish King Valdemar II. In these decades Lübeck became what we call today a boomtown and within this significant upturn the "new" material — brick — found its way into the burgeoning "Hanseatic architecture".

First brick buildings – the end of 12th century

The first buildings made of brick were clerical and official ones. The oldest thermo luminescence dated bricks are from the cathedral's choir, the apse and the transept, dating around 1176 and 1181.[3] It is known from written sources that Duke Henry the Lion, married to Matilda Plantagenet, daughter of King Henry II, arranged the first stone laying of the cathedral in 1178. This date roughly converges with the earliest phase of Lübeck's city walls, having thermoluminescenced dates around 1181, and fits with a mention in the chronicles of Arnold of Lübeck for

---

[1] Schneider 2008, 271.

[2] Harder 2012, Rieger 2012a, 2012b.
[3] Holst 1999, 41.

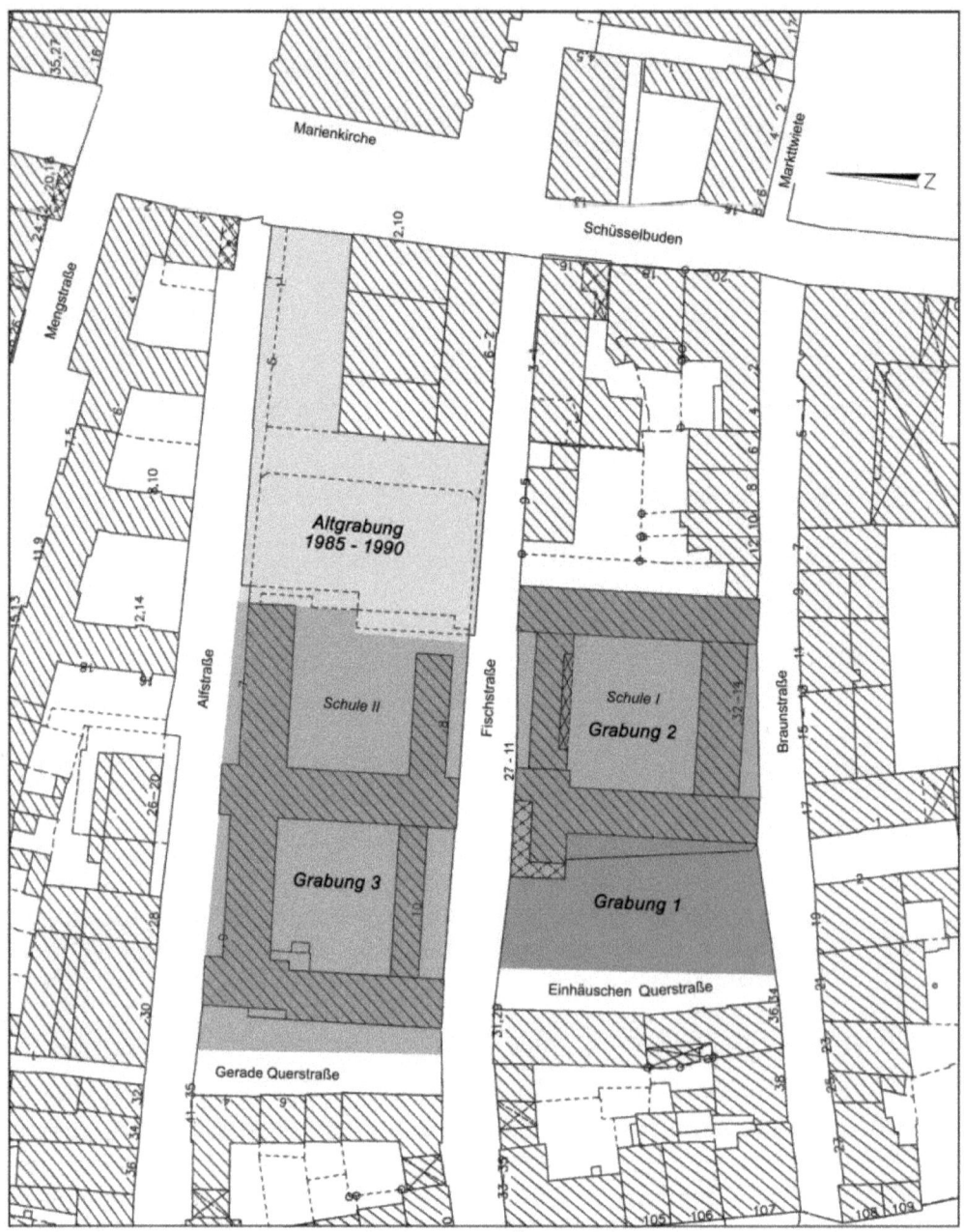

Figure 1. Lübeck. Plan of the excavation Gründungsviertel (plan made by Abteilung für Archäologie der Hansestadt Lübeck).

1179: 'intermiss constructio nibus ecclesiarum cepit firmare poesidia civitatum et urbium, quia bella plurima adversus enim consurgebant'. Not only are the churches mentioned, but also the fortification of Lübeck.[4] Today, remains of this oldest phase are still in situ and visible in the lower parts of the towers flanking the imposing castle-gate (Figure 3).

It is important to point out that the measurements of the earliest bricks in Lübeck are unique. The 12th-century bricks with a height of 7.3–7.8cm, a width of 12.1–12.7cm and a length of 27.8–28.5cm were smaller than the ones in the late medieval and early modern periods. A second phase of early bricks finishing the fortification systems dates to 1217, which is as well confirmed by historical references. These bricks had a height of around 10cm while, curiously, the bricks from the second half of the 13th century onwards (around 1230), had a standardised height of 8.5–9cm.[5] Particularly with regard to the change in brick size and combined with absolute dates from written sources and dendrochronological data, it is possible to think about some sort of brick chronology. Manfred Gläser formulated an initial chronology based on a bundle of criteria concerning the brick sizes, the composition of the clay, the firing, the surface structure and the bond and the joints of the masonry taking from excavations at Untertrave 111/112, Alfstraße

---

[4] Goedecke and Holst 1993, 267.

[5] Gläser 1988, 212.

Figure 2. Lübeck. View on the densely built area of the later excavation before the destructions of the Second World War. Main buildings, wings buildings and square buildings from the 13th–19th century (by Abteilung für Archäologie der Hansestadt Lübeck).

Figure 3. Lübeck. Castle-Gate-Towers with lower levels from the 12th century (picture taken by Abteilung für Archäologie der Hansestadt Lübeck).

36/38, Breite Straße 26 and Johanniskloster.[6] The analysis of seventeen walls from these sites is the base of this Lübeck brick chronology (Figure 4), which — and that has to be emphasised — only works there. It will not be able to date brick walls from other cities in Denmark or Northern Germany by using the Lübeck chronology. We do not know the exact reason, but in other Danish and German cities the brick sizes from one phase of construction could change dramatically.[7]

Early large brick buildings in Lübeck are always associated with the Danish reign, especially under the rule of King Valdemar II, whose politics led on to an alteration and expansion of the harbour to make Lübeck the trade centre of Northern Europe. There are at least three large buildings known to date from the early 13th century, called "Saalgeschoßhäuser". These are large, very prestigious stone (brick) buildings with an oblong, aisled plan form and two fully-furnished storeys. They are often cellared and are frequently found on prominent corners. Two of them are still standing: Koberg 2 and Alfstraße 38 (Figure 5). The house Alfstraße 38 is 12.5 to 24.0m wide and has a 2.2m deep cellar, a 3.1m high first floor and a 3.6m high hall in the second floor.[8] At Koberg 2 the bricks are of the same size as those found in the second phase of the town wall, and

---

[6] Gläser 1987, 245.
[7] Thanks to Morton Søvsø from Ribe, for telling me about the excavation of a 12th century brick building with wide variations in brick size.
[8] Holst 1985, 133.

Fresh Approaches to the Brick Production and Use in the Middle Ages

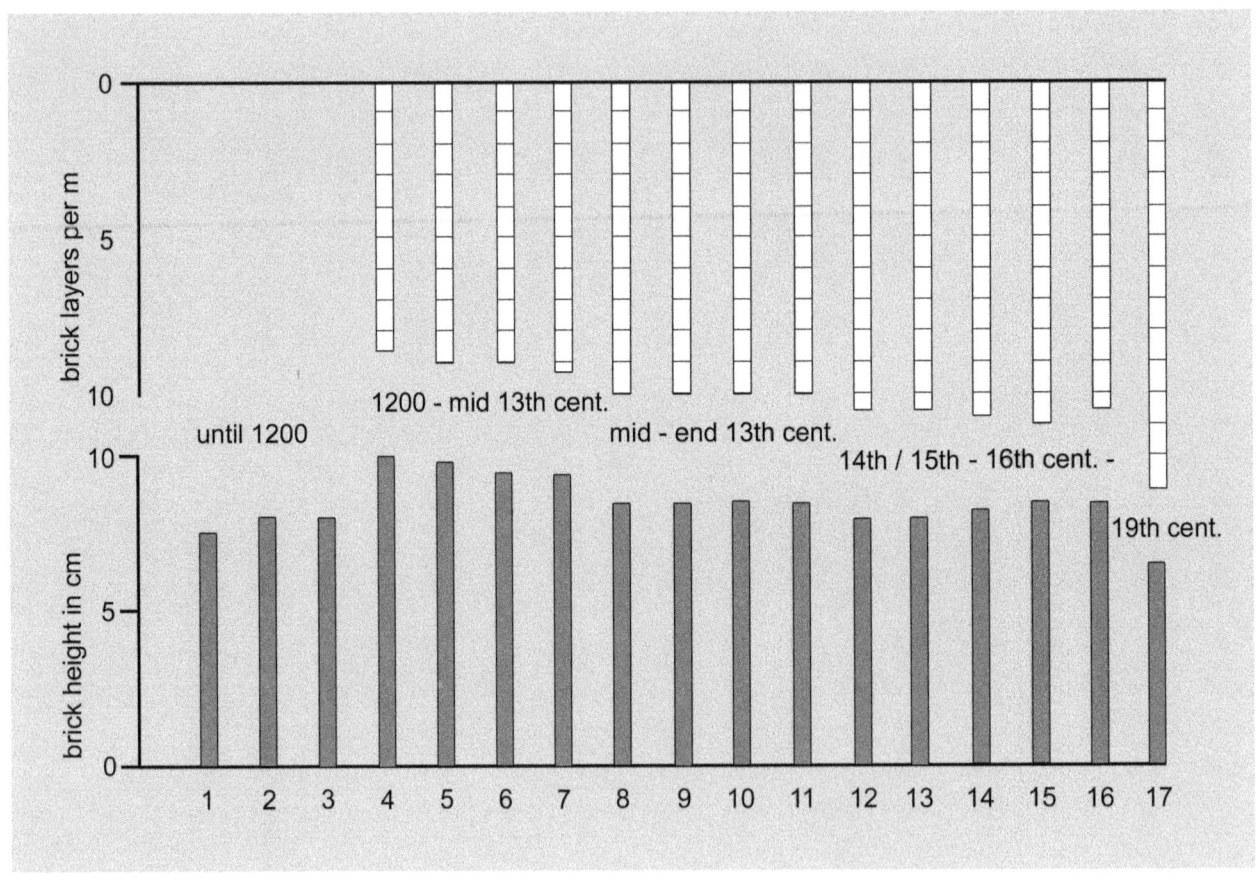

Figure 4. Lübeck. Scheme of alternating brick-sizes during medieval and post medieval periods in Lübeck according to the number of layers built in one meter of masonry (after Gläser 1988).

Figure 5. Lübeck. So-called "Saalgeschoßhaus" at Alfstraße 36, dated 1216 (picture taken by Abteilung für Archäologie der Hansestadt Lübeck).

share the same masonry, bond, technique, mortar and even the special moulded bricks. Ceiling beams from Alfstraße 38 are dendrochronologicallly dated to around 1216, and the thermoluminescenced dates of the bricks at Koberg 2 and Alftstraße 38 are 1216±27.[9] Similar is the excavated Saalgeschoßhaus at Schüsselbuden 6 (Figure 6). It was a large freestanding building on a representative property, and opposite St Mary's church and the market place. The plot was extremely long: 55m long and 27m wide. Late it was divided into six different plots,[10] whose sizes are common in the Hanseatic towns. Unfortunately, only the cellar was left after the destructions of World War II, but this one was unique. It was erected on top of an older 12th century timber building, which could be reconstructed as a three-aisled hall, 8.90m by 16.30m. Interestingly, newer buildings preserved these dimensions and the topographic situation until 1942. This feature, together with Alfstraße 38, rank among the most eminent examples of early brick-architecture in Northern Germany. Its walls were made from the same brick types as at Alfstraße 38, and are dated to the early years after 1200. The cellar had six bays and a groined vault on three brick pillars (Figure 7). The impression of a very representative sales-cellar is still preserved in the neighbouring building Schüsselbuden 2

---

[9] Gläser 1985; Goedecke and Holst 1993, 262, 263; Holst 1985, 134.
[10] Remann 1991, 347.

Figure 6. Lübeck. Saalgeschoßhaus at Schüsselbuden 6 during the excavation in 1985 (picture taken by Abteilung für Archäologie er Hansestadt Lübeck).

Figure 7. Lübeck. Schüsselbuden 6. Excavated brick pillar of the groined vaulting (picture taken by Abteilung für Archäologie der Hansestadt Lübeck).

(Figure 8). The southern wall of Schüsselbuden 6 consists mainly of boulders with a faced brick wall.[11] This part seems to be older than the other walls, whose bond was made of alternating layers characterised by two stretchers and one header (called in Lübeck the Gothic bond), but there were no other traces of this dual wall type of mainly boulders and facing bricks up to the recent excavation. But we will come back to this later.

Small brick buildings

More common in the years around 1200 were smaller brick buildings standing in front or in the middle of the plots, parallel to the streets. They had a more or less square plan, usually with an external staircase, were several stories high and appeared like a tower, and often connected with a timber or brick building at the front, side or back. While in some cases the "Saalgeschoßhäuse" could be seen as kind of guildhalls, the smaller buildings were mostly found in the merchants' area of the Gründungsviertel and seem to be more private.

Figure 8. Lübeck. Schüsselbuden 2. Full in situ standing three-aisled and very representative sales-cellar of the early 13th century (picture taken by Abteilung für Archäologie der Hansestadt Lübeck).

---
[11] Remann 1990, 347; 1991.

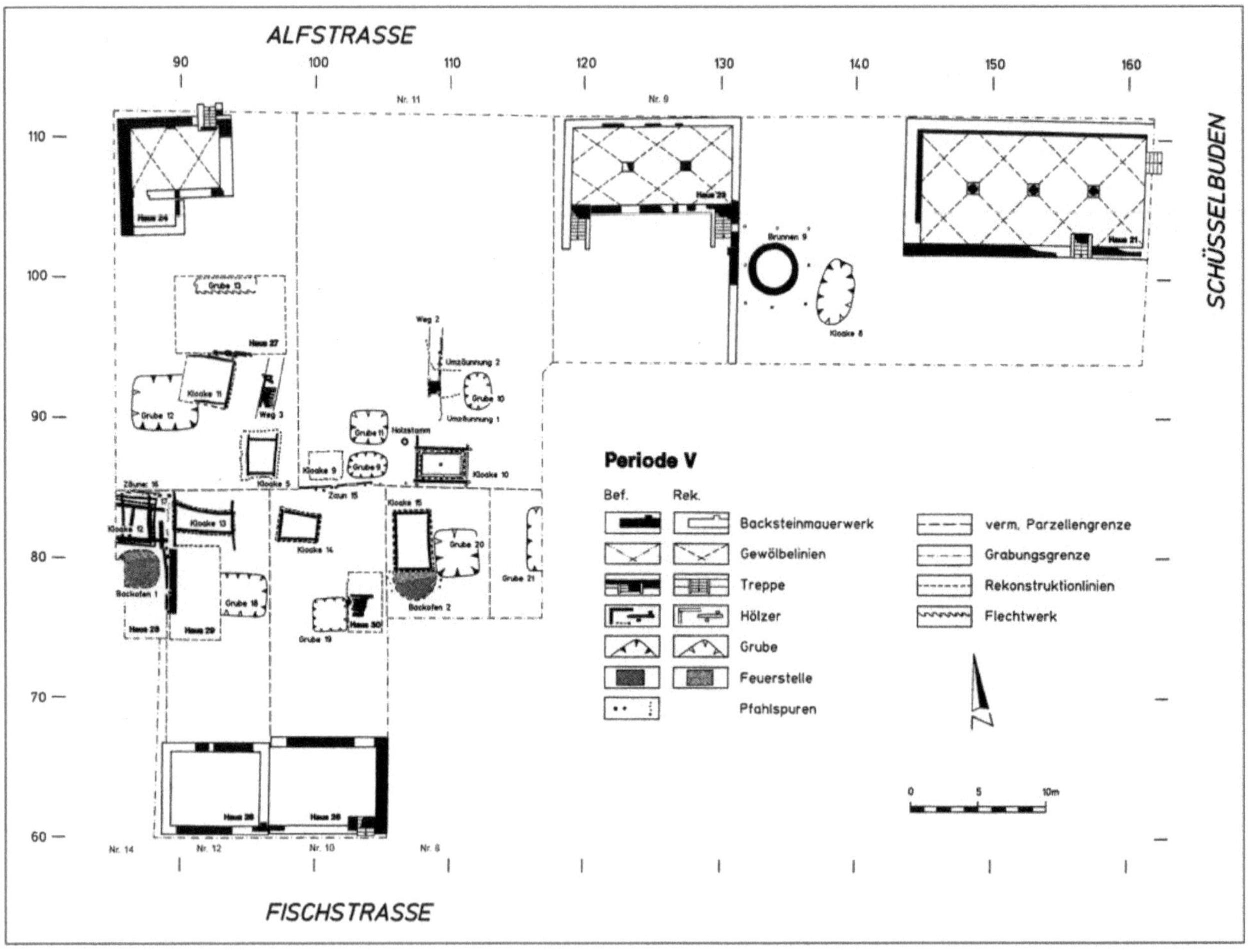

Fresh Approaches to the Brick Production and Use in the Middle Ages

Figure 9. Lübeck. Plan of early brick-phase of the second half of the 13th century. Shown are small brick buildings at Alf- and Fishstraße (after Legant 2010).

During the excavations of the 1980s at least five of these buildings excavated close to the recent project (Figure 9). They were all integrated in the later architecture built after 1250, and had survived in extremely well preserved conditions with sizes of 6–7m and 8–8.5m in length and a height of up to 2.6m. Interestingly, these cellars did not have brick-lined floors — only clay-layers were recorded. Some of the buildings were vaulted and others had only ceiling beams.[12] Further examples from Mengstraße 31 and a free standing building from Fleischhauerstraße 20 complete this study. The 6.2 x 7.8m house from Mengstraße 31 had brick-sizes of 10.1–10.4 cm in height and up to nine layers per meter, and was dated between 1200 and 1230.[13] The 6.1 x 7.3m building from Fleischhauerstraße 20 showed the same construction details and had a wall-thickness of 75cm.[14] The new excavated features from Fischstraße 25 (Figure 10) fit well in this scheme. The front gable was still in situ for more than 2m in height and had remains of a groined vaulting preserved in the masonry.[15] There were two brick pillars showing significant parallels to the previously mentioned Schüsselbuden 6. The entrance had

Figure 10. Lübeck. Fischstraße 25. Remains of a mid-13th century small brick building with two brick pillars and traces of a groined vault in the front-gable of a younger main building (picture taken by Abteilung für Archäologie der Hansestadt Lübeck).

---

[12] Radis 2000, 156.
[13] Radis ip, Schalies 1999, 134.
[14] Fabesch 1989, 140.
[15] Stammwitz 2012b, 56.

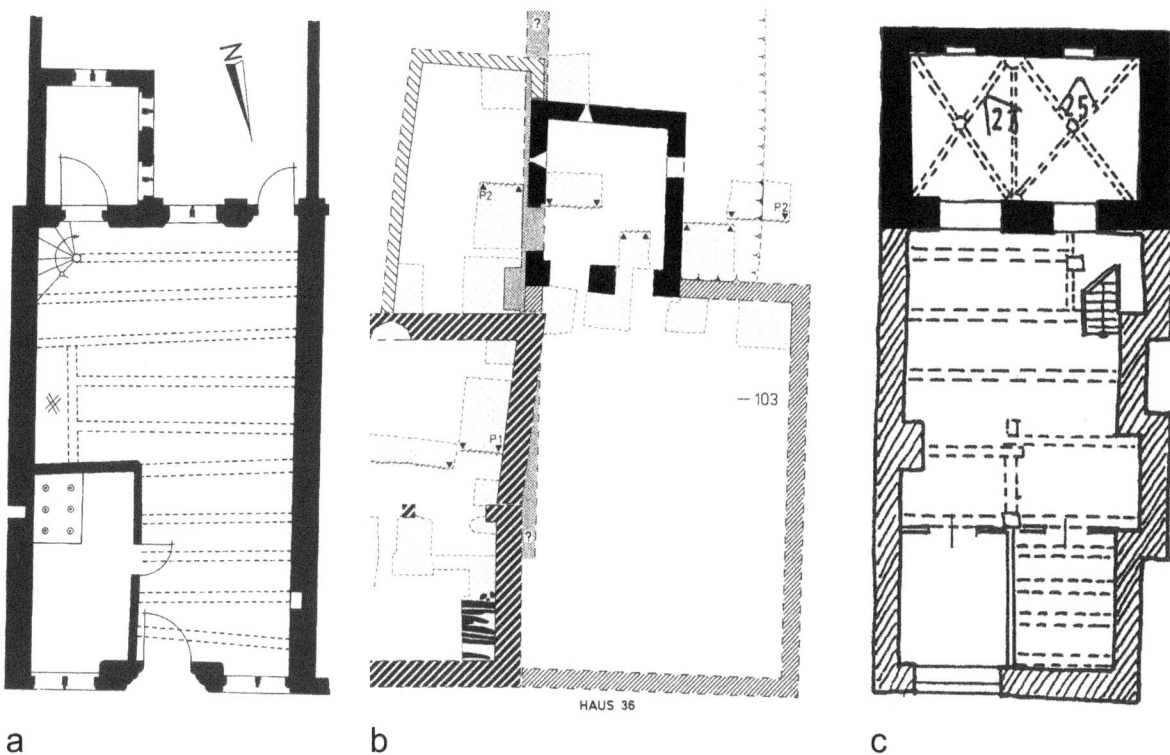

Figure 11. Lübeck. Ground plan of small brick buildings in connection with a Dielenhaus at Hundestraße 94 (a), Alfstraße 36 (b) and Braunstraße 14 (c) (by author).

been at the southern side of the building, leading to the back of the plot. But the brick sizes varied from 8.7 to 9.6cm, so it would be one of the latest examples of these "older" brick-buildings, dating back to the middle of the 13th century. At this time, the so-called "Dielenhäuser" were built: they are large brick-buildings with only one fully developed storey, the lofty ground-floor hall (Diele), after which this house-type was named.

In addition to the Dielenhäuser, the following buildings had been erected in the middle of the properties in a direct connection to the main houses (Figure 11). They were smaller than the older freestanding ones parallel to the streets, and date as well in the second half of the 13th century.[16] Remains of two are still preserved in situ at Hundestraße 94 (Figure 11 a) and Braunstraße 14 (Figure 11 c), while others are excavated at Alfstraße 36 (Figure 11 b), and from the recent project at Fischstraße 19 (Figure 12). The newly-discovered feature was interpreted as the massive foundation for a potential several storey high building, and had a width of approximately 1.40m and a depth of more than 1m. The ground-plan had the same area of around five meters in the square than the other three examples shown in Figure 11. There are lots of interpretations about the function of these kinds of buildings. Some colleagues think about the precursor of the later wing-buildings, some consider them as sorts of fireproof strong-rooms.

Figure 12. Lübeck. Fischstraße 19. Excavation of a foundation of a small brick building standing at the rear gable of the Dielenhaus (picture taken by Abteilung für Archäologie der Hansestadt Lübeck).

To sum up, the brick architecture of the years around 1200 in Lübeck is very diverse and complex. But where are its origins? When did it start in secular architecture? The decades around the middle of the 12th century are dominated by timber-buildings. But when did it change? During our recent excavation we discovered what we call the "missing link" between pure timber and brick-architecture — combined constructed buildings of the second half of the 12th century.

The archaeologically best-preserved main buildings of that period are full-cellared timber constructions made

---

[16] Hammel 1987, 196; Holst 1985, annotation 44; Kruse 1997.

Figure 13. Lübeck. Braunstraße 30. Sill beams of timber-cellar with ramp and wall stringers made of Gotland-limestone, dated 1181 (picture taken by Abteilung für Archäologie der Hansestadt Lübeck).

Figure 14. Lübeck. Fischstraße 21. Brick lined stairs leading to a timber cellar, dating around 1200. The stairs are overbuilt by the rear gable of a mid-13th century Dielenhaus (picture taken by Abteilung für Archäologie der Hansestadt Lübeck).

in a frame technique with sills, posts and wall-planks. They, therefore, are documented in a large number of archaeological reports on Lübeck.[17] Recently we were able to excavate the remains of a 5 x 6m wide timber-cellar at Braunstraße 30 with unique construction details. The complex is dendrochronologically dated to 1181 and — except to all the oldest features — combined with another building material: limestone. Not only were the beams underpinned with plates of limestone, but also the whole entrance including the 21° steep ramp and the wall stringers were constructed in dry masonry wall with clay mortar (Figure 13). After analysing the limestone we believe that it came from Gotland because of the very specific composition of corals that are common there.[18] In addition, written sources mention the trade relations between Gotland and Lübeck in 1161, so it is easily conceivable the stone coming as a kind of ship-ballast to Lübeck, or as a basic ingredient for making mortar in respect of the builders' workshop of the brick built churches or the city wall. The source tells us about Henry the Lion, Duke of Saxony and Bavaria, registering the peace between Germans and Gotlanders and confirming the libertines given to them by Emperor Luther.[19] Twenty years later in 1181, and close to the harbour area, it is hardly surprising that there was a chance to get such a material for the use of house building. Nevertheless, this feature was the first recorded combination of the two different building materials and was unique in Lübeck. But a few weeks later, evidence of this "missing-link" was confirmed twice.

The next example was excavated at the middle of plot at Fischstraße 21. It is a three-part complex consisting of a timber-cellar inside a potentially square brick building with a lateral brick lined staircase (Figure 14 and 15).

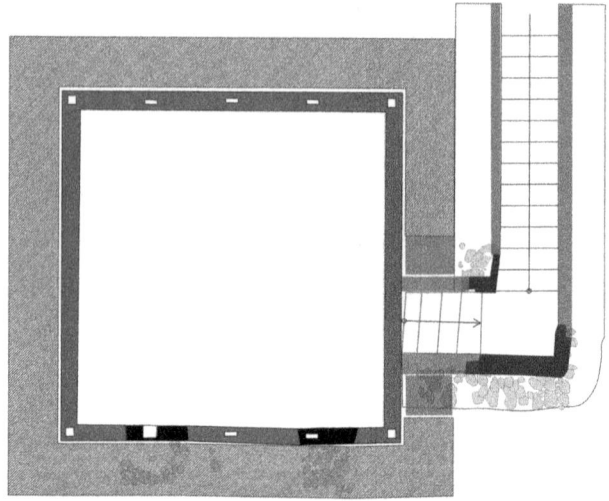

Figure 15. Lübeck. Reconstruction of a small brick building with timber cellar and brick lines stairs excavated at Fischstraße 21 (plan made by Abteilung für Archäologie der Hansestadt Lübeck).

All features of this complex were badly damaged by the rear-gable of a mid-13th century Dielenhaus, as well as the construction pit of the modern school cellar of the 1960s. The build-up technique of the wall-stringers is already known from the example of Schüselbuden 10 mentioned above. We recorded boulders behind a cladding of 7cm high bricks set in Gothic bond and in lime mortar. The bricks had exactly the same characteristics as the other oldest bricks from Lübeck. In comparison, the bricks used in the steps had a different size and a height of 10cm. According to the Lübeck chronology, and in respect of the youngest features

---

[17] Fehring 1989; Gläser 2001; Legant 2010; Müller 1992, 146f; Schalies 1999, 126f; 2003, 34.
[18] Thanks to Ursula Radis for giving me information about the analyses.
[19] Sprandel 1982, 173ff.

Figure 16. Lübeck. Fischstraße 21. Large timber cellar with brick-lined stairway and two dividing inner walls. The brick wall behind the cellar was already taken away in the state of record (picture taken by Abteilung für Archäologie der Hansestadt Lübeck).

Figure 17. Lübeck. Fischstraße 17. Construction diagram and detail photo of the timber cellar with sills, posts, wall planks, top plate and ceiling beam (picture taken by Abteilung für Archäologie der Hansestadt Lübeck).

(the steps), this complex has to date to the years around 1200. Unfortunately the dendrochronological analysis of the timber cellar sill beams dates around or after 1155, so there is no exact date, but around 1200.[20] However, what is also important to point out is that this combined timber-and-brick-building seems to be part of a square, maybe a tower-like building which is a common house-type in Westfalia and lower Saxony in this period of time and is called in German "Steinwerk" or in Latin "domus lapidea".[21] The surrounding boulders functioned as a kind of foundation for the rising walls. But to be sure about this more excavation work has to be done in the next few weeks to verify this thesis. Convergent brick-buildings, standing in the front or the middle of the properties, were already mentioned for Johanniskloster, Fleischhauerstraße 20, Mengstraße 31, Alfstraße 36 and known from Julius-Leber-Straße 36.[22]

The third example is another where a large timber cellar and a brick lined stairway connected with a brick lined external wall are combined (Figure 16). The building could be seen as an annex to a main timber building and stood in the middle of the property.[23] The architectural construction of timber cellars in Lübeck is in most cases always the same (Figure 17). There are sill beams with posts, wall planks, top plates and ceiling beams joined together without a single nail or bolt. Only differentiated elaborations, special designed notches, function as connecting elements like halving joints with square mortises held together by posts with tongues and wall planks standing on rabbets being held in position by the earthen pressure of the building pit. But the most interesting aspect of this complex is the combined construction of the 9 x 5m timber cellar and the

Figure 18. Lübeck. Fischstraße 21. Brick-lined stairway to timber cellar with Gothic bond and remains of barrel vault set against the door-lintel (picture taken by Abteilung für Archäologie der Hansestadt Lübeck).

brick-lined features. The stairway consists of the known brick material from the end of the 12th century; it has the same type of structure as the stairway in the second example, or the southern wall from Schüsselbuden 10. In this case the bricks in the step also have the special height of around 7cm. Yet there are no traces of it being an additional feature, so the stairway and its elements connect direct to the posts and the lintel of the doorway (Figure 18). Its brick lined wall-cladding has specially shaped bricks with a diagonal bearing surfaces for a small barrel vault, which once spanned the whole staircase. It led from a timber main building into the cellar and must have looked

---

[20] Stammwitz 2012.
[21] Rieger 2010, 117ff.
[22] Fabesch 1989, 140ff; Gläser 1989, 15; Holst 1985, 137ff ; Radis (in progress); Remann (in progress); Wiedenau 1983, 156.
[23] Radis 2013.

very impressive to a merchant's potential customer — we suppose a merchant to have lived and worked here in the late 12th century. The dendrochronological analyses date the timber in the years around 1180, so does the brick chronology. Similar bricks of the same sizes were used to build up a wall behind the cellar in the south (Figure 19). Unfortunately, it had mostly been destroyed by newer cesspits and pits of the 13th century, but we know that the wall only reached from end-to-end of the southern cellar and was erected and in use at the same time. There were no traces of the same wall on the eastern or western sides. Constructively it was made in the same technique as already mentioned for other 12th century walls in Lübeck with boulders behind a brick lined cladding in Gothic bond. This wall may have just been a foundation for a sill beam, just elevated for protection against rain or soil moisture. On the other hand it could have been a ground floor wall covering the rear-gable of the annex building. Maybe it was a part of a potential kitchen or a workshop needing to be separated from the timber beams. But these theories need more research. Nevertheless, this is hitherto the oldest dated example of secular brick lined architecture in Lübeck. It also confirms the dating of around 7cm high bricks. In addition, it is one of three examples presented here, which provide the link between pure timber architecture from the mid-12th century and the brick architecture of the 13th century. This excavation is still on going, and we hope that there will be more discoveries to prove our theories and to give new answers about the beginnings of the secular brick buildings on the Baltic Sea.

Figure 19. Lübeck. Fischstraße 17. Detail of a brick wall behind the timber cellar (here the south western corner post with top plate and the ceiling beam) (picture taken by Abteilung für Archäologie der Hansestadt Lübeck).

Conclusions

The recent excavations in the Gründungsviertel in Lübeck revealed new features concerning secular brick architecture of the late 12th and early 13th century. Long searched missing links between pure timber- and brick-architectures could be excavated in several examples. Here special sized bricks of a height of 7 cm were used in walls and especially wall-stringers and stairways leading into late 12th century timber cellars. These combinations of two different building materials were the first that had been discovered in Lübeck. As well as due to TL-dating the bricks fit well in the brick-chronology of the Hanseatic Town, which is proven by new dendro dates. Only a few years later when Lübeck became Danish, after the takeover of King Valdemar II, brick was the number one element to erect houses. Then smaller brick buildings or even huge halls rose along the streets in a rectangular or even square shape. Timber buildings were only built as outbuildings. Again bricks of these decades had a special measurement, in that case of about 10cm heights. Leading over to the large Dielenhäuser of the 2nd half of the 13th century the brick development almost came to a standardised moulded pattern of a height of 8.5cm. Bricks could be seen as the most important building material in the later capital of the Hanseatic League that first was used to erect official and clerical architecture. But a little while later it was also the exclusive building substance to erect the prestigious constructions of rich merchants alongside the main streets of the oldest quarter in Lübeck.

References

Fabesch, U. H. 1989. Archäologische und baugeschichtliche Untersuchungen in der Fleischhauerstraße 20 zu Lübeck. In Lübecker Schriften zur Archäologie und Kulturgeschichte 16, 137–159, Bonn, Dr. Rudolf Habelt.

Gläser, M. 1985. Befunde zur Hafenrandbebauung Lübecks als Niederschlag der Stadtentwicklung im 12. und 13. Jahrhundert. Vorbericht zu den Grabungen Alfstraße 36/38 und Untertrave 111/112. In Lübecker Schriften zur Archäologie und Kulturgeschichte 11, 117–129. Bonn, Dr. Rudolf Habelt.

Gläser, M. 1987. Archäologische Beiträge zur Datierung der Lübecker Backsteinmauern. In Archäologisches Korrespondenzblatt 17. 245–252.

Gläser, M. 1988. Die Lübecker Backsteinchronologie. In Lübecker Schriften zur Archäologie und Kulturgeschichte 17, 210–212. Bonn, Dr. Rudolf Habelt.

Gläser, M. 1989. Archäologische und bauhistorische Untersuchungen im St. Johanniskloster zu Lübeck. Auswertung der Befunde und Funde. In Lübecker Schriften zur Archäologie und Kulturgeschichte 16, 9–120. Bonn, Dr. Rudolf Habelt.

Gläser, M. 2001. Archäologisch erfasste mittelalterliche Holzbauten in Lübeck. In M. Gläser (ed.), Lübecker Kolloquium zur Stadtarchäologie im Hanseraum III. Der Hausbau, 277–305. Lübeck, Schmidt-Römhild.

Goedicke, Ch. and Holst, J. Ch. 1993. Thermolumineszenzdatierung an Lübecker Backsteinbauten. Probleme und Entwicklungen. In R. Hammel-Kiesow (ed.), Wege zur Erforschung städtischer Häuser und Höfe, 251–271. Neumünster, Wacholtz Verlag GmbH.

Hammel, R. 1987. Hauseigentum im spätmittelalterlichen Lübeck. Methoden zur sozial- und wirtschaftsgeschichtlichen Auswertung der Lübecker Oberstadtbuchregesten. In Lübecker Schriften zur Archäologie und Kulturgeschichte 10, 85–300. Bonn, Dr. Rudolf Habelt.

Harder, J. 2012. Hölzerne Infrastruktur des Mittelalters aus dem sogenannten Gründungsviertel der Hansesatdt Lübeck. In Mitteilungen der Deutschen Gesellschaft für Archäologie des Mittelalters und der Neuzeit 24, 123–130. Paderborn, Deutsche Gesellschaft für Archäologie des Mittelalters und der Neuzeit.

Holst, J. Ch. 1985. Zur Baugeschichte der Häuser Alfstraße 36 und 38 in Lübeck. Ein Zwischenbericht. In Lübecker Schriften zur Archäologie und Kulturgeschichte 11, 131–143. Bonn, Dr. Rudolf Habelt.

Holst, J. Ch. 1999. Dar umme is se noch so ordenliken buwet – Früher Backsteinbau in Lübeck. In Schriften des Instituts für Bau- und Kunstgeschichte der Universität Hannover, Band 12, 41–51. Hannover, Hannover.

Kruse, K. B. 1982. Zu Untersuchungs- und Datierungsmethoden mittelalterlicher Backsteinbauten im Ostseeraum. In Archäologisches Korrespondenzblatt 12, 555–562.

Kruse, K. B. 1997. Die Baugeschichgte des Heiligen-Geist-Hospitals zu Lübeck. In Lübecker Schriften zu Archäologie und Kulturgeschichte 25. Bonn, Dr. Rudolf Habelt.

Legant, G. 2010. Zur Siedlungsgeschichte des ehemaligen Lübecker Kaufleuteviertels im 12. und frühen 13. Jahrhundert. In Lübecker Schriften zu Archäologie und Kulturgeschichte 27. Rahden/Westfalen, VML.

Radis, U. ip. Zur Siedlungsgeschichte des Lübecker „Kaufleuteviertels" vom 13. bis zum 20. Jahrhundert. Die Befunde der Steinbauperioden der Grabung Alfstraße / Fischstraße (HL-70). In Lübecker Schriften zur Archäologie und Kulturgeschichte. In progress.

Radis, U. 2000. Die Turmhäuser der Grabung Alfstraße/Fischstraße in Lübeck und ihre Datierungsmöglichkeiten. In D. Schuhmann (ed.), Bauforschung und Archäologie im Spiegel der Baustrukturen, 149–161. Berlin, Lukas.

Radis, U. 2003. Ein Holzhaus in 8 Meter Tiefe. In M. Gläser, D. Mührenberg (eds), Weltkulturerbe Lübeck. Ein Archäologischer Rundgang, 14–15. Lübeck, Schmidt-Römhild.

Radis, U. 2013. Ein Holzkeller der besonderen Art. In Archäologie in Deutschland 2/2013.

Remann, M. ip. Jüngere Steinbebauung im Ostteil der Grabung Alfstraße-Fischstraße-Schüsselbuden aus den Jahren 1985–1990. In Lübecker Schriften zur Archäologie und Kulturgeschichte. In progress.

Remann, M. 1990. Die ersten steinernen Bürgerhäuser zu Füßen der Marienkirche. In Die Heimat 12 / 97, 347–348.

Remann, M. 1991. Romanische Backsteinbebauung im Zentrum von Lübeck. Ein Beispiel zu Füßen der Marienkirche. In Arbeitskreis für Hausforschung (eds), Berichte zur Haus- und Bauforschung 1, 9–16. Marburg, Jonas.

Rieger, D. 2010. platea finalis. Forschungen zur Braunschweiger Altstadt im Mittelalter. Rahden/Westfalen, VML.

Rieger, D. 2012a. Zeitliche Tendenzen und Konstruktionskontinuitäten – aktuelle Befunde zur Holzarchitektur der Großgrabung im Lübecker Gründungsviertel. In Mitteilungen der Deutschen Gesellschaft für Archäologie des Mittelalters und der Neuzeit 24, 131–140. Paderborn, Deutsche Gesellschaft für Archäologie des Mittelalters und der Neuzeit.

Rieger, D. 2012b. Recent excavations in Lübeck – Slavonic traditions and the beginning of the capital of the Hanseatic League. In MERC – EAA 2012, Helsinki. In progress.

Schalies, I. 1992. Archäologische Untersuchungen zum Hafen Lübecks. Befunde und Funde der Grabung an der Untertrave/Kaimauer. In Lübecker Schriften zur Archäologie und Kulturgeschichte 18, 305–344. Bonn, Dr. Rudolf-Habelt.

Schalies, I. 1999. Neue Befunde hochmittelalterlicher Holzbauten im Lübecker Gründungsviertel. In Archäologisches Korrespondenzblatt 29, 125–141.

Schneider, M. 2008. Archäologie im Lübecker Gründungsviertel – Fragestellungen, Chancen und Perspektiven neuer Großgrabungen. In F. Biermann, U. Müller and Th. Terberger (eds), „Die Dinge beobachten…" Archäologische und historische Forschungen zur frühen Geschichte Mittel- und Nordeuropas. Festschrift für Günter Mangelsdorf zum 60. Geburtstag, 271–282. Rahden / Westfalen, VML.

Schneider, M. 2009. Lübeck im 12. und 13. Jahrhundert. Archäologische Befunde zur Erntstehung einer mittelalterlichen Grossstadt. In Expansion – Integration? Danish – Baltic contacts 1147–1410 AD, 75–93. Vordingborg, Danmarks Borgcenter.

Sprandel, R. 1982. Quellen zur Hanse-Geschichte. In Ausgewählte Quellen zur deutschen Geschichte des Mittelalters 36, 173–175. Darmstadt, Wissenschaftliche Buchgesellschaft.

Stammwitz, U. 2012a. Für gehobene Ansprüche. In Archäologie in Deutschland 1/2012, 54.

Stammwitz, U. 2012b. In bester Lage. In Archäologie in Deutschland 3/2012, 56.

Weidenau, A. 1983. Katalog der romanischen Wohnbauten in westdeutschen Städten und Siedlungen. Tübingen, Wasmuth.

The Origin and Rise of Brick Technology and Use in Medieval Pomerania, Pomerelia and Lower Silesia

Felix Biermann and Christofer Herrmann

Abstract: The first brick buildings in Pomerania, Pomerelia (modern North-Eastern Germany and North Poland) and Lower Silesia (Western Poland) date from the late 12th century. Since the 13th century this building material became prevailing in the ambitious architecture. An important role for this innovation and the increase of the new building material was played by the monasteries which were founded in Pomerania since the 1150s, as well as some important castles with large brick towers which were established in that region in the decades shortly after 1200. Danish influence seems to have been important for the mediation of the building material and technique to Pomerania. Archaeological, building- and art-historical research in Pomeranian abbeys (Stolpe/Peene, Grobe, Belbuck, Kolbatz/Kołbacz, Eldena, Bergen, Altentreptow, Oliva/Oliwa and other), further also from castles (in particular Haus Demmin, Stolpe/Oder) and village churches (for instance Altenkirchen on Rügen) deliver interesting information for understanding the chronology, the initial points and the directions of the technical-stylistic influences and the origin of brick in this region. A similar situation can be observed in Lower Silesia, where the earliest brick buildings are found in Cistercian monasteries (Leubus/Lubiąż, Trebnitz/Trzebnica). This paper gives an overview of the research. Also the archaeological evidence for the production of brick and the artistic-architectural design with bricks are stressed as central topics.

Keywords: brick architecture, castles, Danish influences, Lower Silesia, Middle Ages, monasteries, Pomerania, Pomerelia

Introduction

Until the 12th century Pomerania, the country by the sea on today's north-east German and northern Polish coast with the island of Rugia, was a still widely pagan Slavic area. Since the 1120s this situation changed as mission efforts increased, and by the end of the century the bulk of the people living in this area had converted to Christianity[1]. Well-known events in this process were the mission trips of Bishop Otto of Bamberg 1124/25 and 1128[2] and the destruction of the important temple stronghold on Cape Arkona on the island of Rugia by the Danes in the year 1168[3]. At the same time three mighty powers were rising in the region – the Greif dukes in Pomerania on both sides of the river Oder, the Samborid dukes in Pomerelia and the princes of Rugia on the island of the same name. The western parts of Pomerania became a part of the Holy Roman Empire; from the 1160s to the 1220s Danish Valdemarian kings were able to get control over this territory[4]. Since the late 12th century, increasingly since the 1220s, the immigration of western population in these territories started, in the course of the German eastern settlement[5]. This migration and acculturation process led to a "westernisation"[6] of the country. Its culture changed in many aspects.

Already Bishop Otto erected wattle and daub churches in the castle-towns he visited[7], and he established the basis to found a Pomeranian bishopric in Wollin/Wolin, which was carried out in Cammin/Kamień Pomorskie[8] in 1140. In the following decades also monasteries were donated – in the Pomeranian Greif duchy since the 1150s and on the island of Rugia since the 1190s. This time was the beginning of stone architecture, which was revolutionary in a region where up to then the building culture had been characterized exclusively by wood, wickerwork and clay. Soon afterwards also the brick building techniques emerged which still today characterize Pomerania in the form of the magnificent Gothic brick architecture.

Pomerelia (the territory west of the mouth of river Vistula), the eastern neighbor of Pomerania, was already christianized in the late 10th century by the Polish Piast dynasty. During the mid-12th Pomerelia became an independent Slavic dukedom under the power of the Samborides. Not until this time the country got a basic Christian substructure and a first monastery in Oliva was founded in the 1180s.

Until the 9th and 12th centuries the territory of Lower Silesia was characterized by a tribal structure typical for East Central Europe at that time. During the 9th/10th century Silesia belonged to Great Moravia and from the 10th century onwards the country was christianized gradually. Since the 980/90s the Piastic ruler Mieszko I

---
[1] See for an overview Rębkowski 2011.
[2] Petersohn 1979, 213–232.
[3] Ruchhöft 2010.
[4] For an overview of Pomerania and Rugia see: Benl 1999, 37–9, 100–2; for Pomerelia Quandt 1856.
[5] Benl 1999, 48–85.
[6] Hardt 2005, 17.

[7] Biermann 2006.
[8] Petersohn 1979, 262–341.

Figure 1. Stolpe upon Peene, Benedictine Abbey, view on the excavated church and the ruin of the westwork from the East, 12th century (photo by F. Biermann).

Figure 2. Grobe on Usedom, Praemonstratensian monastery, relics of the northern wall of the church with fieldstone ashlars, 12th century, damaged by late medieval graves (photo by F. Biermann).

succeeded with assuming power in Silesia. From then on this territory was under the rule of the Piast dynasty[9].

In the 12th century the Polish Empire was split into princedoms under different sub-lines of the Piasts which were only loosely intertwined, amongst them the principality of Silesia. Since the 1160s the Silesian Piasts intensified their connections with the Holy Roman Empire. In particular the German immigration led to changes, mainly in the 13th century. But brick architecture appeared for the first time already in the late 12th century by mediation of the Cistercian order (Heinrichau/Henryków).

The focal point of this paper is the dating and origin of brick technology in Pomerania, Pomerelia and Lower Silesia and the role of the monasteries in this innovation, in the light of new archaeological and art historical investigations.

The beginning of brick building and the role of monasteries

The first monasteries in Pomerania were established in the 1150s, with Stolpe upon Peene and Grobe on the island of Usedom. Both monasteries, donated by Greif dukes, are in large part lost today because of their abolishment in 14th/16th century, but excavations and building studies deliver some information on their architecture. At Stolpe, founded as a Benedictine convent in 1153, with monks from Berge near Magdeburg, was built shortly after the donation a three-naved basilica without transept of approximately 39m in length and 13m in width (Figure 1). The church had a straight, closed choir and a mighty west building, which has survived as a ruin up to today. Ground plan and ruins show stylistic influence from Magdeburg, which is no wonder given the provenience of the monks and the large importance of the Church in Magdeburg for this monastery.

The building was made by erratic blocks and ashlars[10]. In the Praemonstratensian monastery of Grobe, founded in the suburbs of the important central castle of Usedom around 1155, a single-nave hall church of at least 33m length and 10m width was erected in the 1150s. The exact length and the eastern architecture of the building are unclear because archaeological and geophysical research brought no clear results as the area has been heavily destroyed by erosion and agrarian works. Also, this building was not made of bricks but of fieldstone and ashlars (Figure 2)[11]. This is also the case with the first Pomeranian cathedral in Cammin: the oldest phase of stone construction (Figure 3) – there was probably a wooden precursor – shows ashlars, and probably not before 1200 the building was continued with bricks[12].

It seems that in the middle and third quarter of the 12th century, in the early times of monastery founding, brick buildings were not yet common. The church of the Cistercian nunnery of Seehausen south of Prenzlau in the Uckermark, established probably by the Pomeranian dukes in the first third of the 13th century, was still built of erratic blocks or fieldstones. After its abolishment in the period of Reformation it disappeared from the ground, but archaeological and geophysical investigations delivered information on its architecture. The excavations in the church area revealed the basements of a single-naved hall of 40–42m length and 13m width, constructed apparently without bricks. Only the cloister to the south of the church, according to the excavations with building phases from the 13th to the 16th century, was built of bricks on boulder packing[13].

The first brick buildings which we know in Pomerania

---

[9] See Grünhagen 1884; Grodecki 1933; Appelt 1971; Moździoch 1990.
[10] Fait 1957/58; Fait 1963; Zaske 1969; Szczesiak 2005, 401–2, Figure 2; Rębkowski 2011, 32–4.
[11] Biermann et al. 2013a.
[12] See for the building history of the Cammin cathedral Rębkowski 2011, 31 p., with further literature.
[13] Investigations 2011/2012. See Biermann et al. 2013b.

Figure 3. Cammin/Kamien Pomorskie, the fieldstone ashlar portal in the northern transept, probably late 12th century (photo by F. Biermann).

originate from the late 12th century. One of the oldest buildings still standing – probably the oldest of all – is the church of Bergen on the island of Rugia (Figure 4), founded by Prince Jaromar I about 1180 as a castle-town church and handed over to Benedictine nuns in 1193; choir, transept and west tower front of the still existing cruciform basilica go back to the late 12th century (1180–1200). It is constructed in brick masonry on a fieldstone basement, supposedly by Danish masters[14]. Another important example is the village church of Altenkirchen on Rugia, not far away from the Slavic Arkona stronghold. The oldest preserved part of the church, the semicircular apse (Figure 5), shows still Romanesque characteristics, with a crossed triangle frieze on human mask-consoles in combination with a double tooth-frieze. This building shows strong Danish influence. From its architecture and style the apse is dated shortly before or around 1200[15]. Dendrochronological data from the roof framework, delivering dates between the 1250s and the 1260s, gave reason to redate its construction to the 1250s[16]. The research has not yet been fully published, but this basis seems not convincing to redate the building contrary to the art historic analysis: There is no proof at all that the tree-ring dated beams belong to the first roof construction. Furthermore, parts of the Cistercian monastery churches of Dargun[17] and Eldena are dated to the early 13th century[18]. The recently excavated church of the nun's convent at Altentreptow, where building works started between 1193 and 1204 (but were not completed), was probably built in brick, too, according to the brick debris in the bedding trenches. But this result is not absolutely sure: in situ verifiable was only the fieldstone packing of the semicircular apse of the single-naved church[19].

Figure 4. Bergen on Rugia, church of the benedictine's nunnery, eastern part and transept, late 12th/13th century (photo by F. Biermann).

Figure 5. Altenkirchen on Rügen, parish church, apse, late 12th/13th century (photo by F. Biermann).

---

[14] Holst 2005; Schäfer 2005; Reimann et al. 2011, 78.
[15] Feldmann 2000, 4–6.
[16] For the dendrochronological dating: Seip Bartholin 1995; Heußner 2005, 126, Figure 2; for the interpretation and redating: Ruchhöft 2010, 78, Figure 98; Reimann et al. 2011, 79; Büttner 2007, 66.
[17] Brachmann et al. 2003.
[18] Kloer 1929; Mangelsdorf 1999.
[19] Biermann 2012.

In Pomerania east of the river Oder the earliest brick buildings are situated in Cammin (cathedral) and Kolbatz (Cistercian abbey). Further to the east the first use of brick took place at the Cistercian abbey in Oliva/Oliwa in Pomerelia near the mouth of the river Vistula. Both monasteries were connected with the Danish Cistercians: Kolbatz was a daughter house of Esrom on Zealand and Oliva a daughter house of Kolbatz. Bogislaw I, Duke of Pomerania, founded Kolbatz[20] in 1173, and one year later the first group of monks from Esrom arrived there. According to the „Annales Colbacenses", the building of the nowadays existing church made of brick was started in 1210. Therefore, there must have been a long-lasting temporary predecessor building in this place, possibly also made of brick. During the first building phase (lasting until the middle of the 13th century) the Cistercians erected the chancel (later replaced by a gothic one), the transept and the eastern part of the nave. Many details show influence of the Zealand brick building tradition and verify the participation of Danish craftsmen (Figure 6).

In 1186 first group of monks was sent 300km east to Oliva[21], where the Duke of Pomerelia, Subislaw I, had founded a new abbey as a burial place for his dynasty (Figure 7). Its building history is very complicated and far from being clarified in detail. There were many building phases and interruptions and we know no documents providing information about the beginning of the building works. Since the Danish elements are less strong than in Kolbatz, we can conclude that in Oliva the building works started later, probably some years after 1210. Some details suggest that apart from Danish bricklayers also craftsmen from Northern Germany were hired to build the monastery. Oliva is an impressive example of technology transfer organised by the Cistercian order. Looking from Oliva, the nearest building places with brick use in the early 13th century were many hundred kilometres away: in Kolbatz and Cammin (300km to the west), Riga (500km to the northeast), Leubus/Lubiąż and Trebnitz/Trzebnica in Lower Silesia (more than 500 km to the south-west).

A situation similar to Pomerania we can notice in Lower Silesia. The first brick building in this region was the Cistercian church in Leubus[22], founded in 1163 and inhabited by the first monks in 1175. The oldest known church, the building of which probably started at the end of the 12th century, was a relatively small building with a typical Cistercian ground plan but in a reduced version (only one side chapel on each aisle of the transept). The church was later completely replaced by a larger gothic building (Figure 8). Based on archaeological research, we know only the ground plan of this first Silesian brick building and no details of the upper construction. A very important and still existing early monumental architecture made of brick in Lower Silesia is the church of the Cistercian nuns

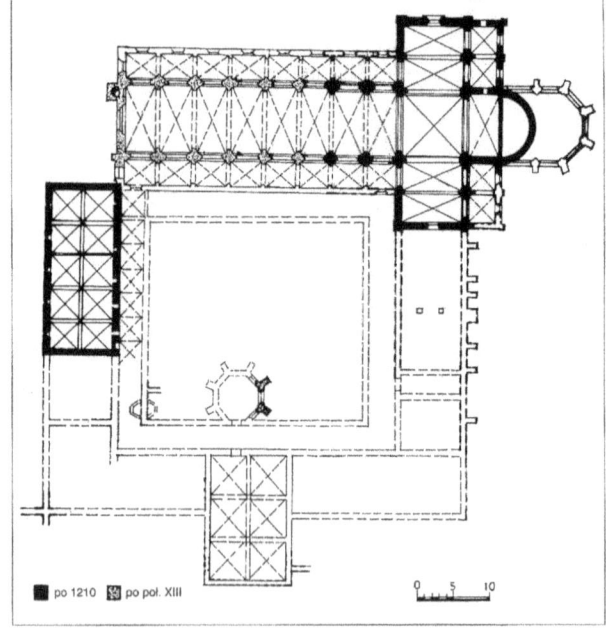

Figure 6. Kolbatz/Kołbacz, Cistercian abbey. Above: view of the church from north-west (photo by C. Herrmann); below: ground plan after Świechowski 2009.

Figure 7. Oliva/Oliwa, Cistercian church: view from south east (photo by C. Herrmann).

---

[20] Krongaard Kristensen 2009, 62–67; Świechowski 2009, 164–172; Nawrocki 2010, 202–206.
[21] Świechowski 2009, 331–340, with further literature.
[22] Świechowski 2009, 278–282, with further literature.

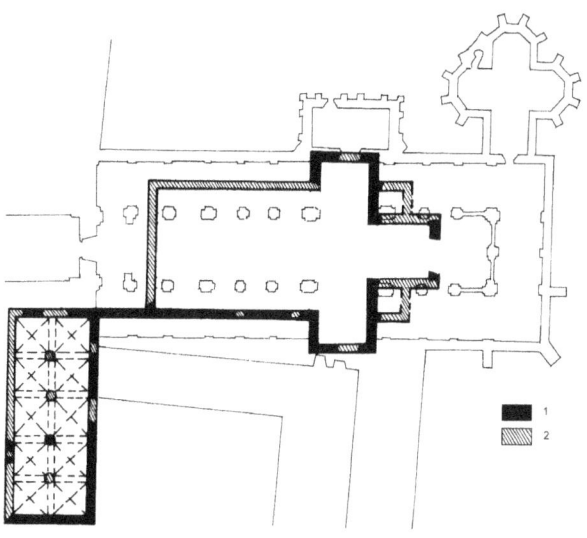

Figure 8. Leubus/Lubiąż, Cistercian abbey: ground plan after Świechowski 2009.

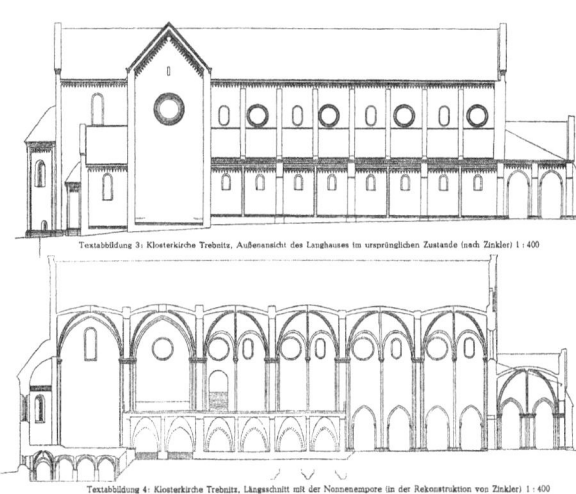

Figure 9. Trebnitz/Trzebnica, church of Cistercian nuns: northern elevation and longitudinal section (plans after Tintelnot 1951).

in Trebnitz/Trzebnica (Figure 9)[23]. It was erected between 1203 and 1240 by initiative of Duke Henry I the Bearded and his wife Saint Hedwig of Silesia with the help of the Cistercians in Leubus. In the same period (about 1220/30) Duke Henry I started to build his impressive residence in Liegnitz/Legnica[24], also made of brick (Figure 10). One of the daughter houses of Leubus, the Cistercian abbey in Mogiła near Cracow, was the first big brick building in Lower Poland (erected from about 1230 onwards).

As a conclusion we may say that between the 1180/90s and the beginning of the 13th century the first brick buildings were established in Pomerania, Pomerelia and Lower Silesia, initially parallel to fieldstone or ashlar masonry. The earliest examples of brick use come from the ecclesiastical-monastic context. The important role of the monasteries for the brick technology is evident also from the fact that most archaeologically documented medieval brick ovens in Pomerania – albeit not before the 13th century – belong to the monastic context: This is the case with a brick kiln at Wackerow, belonging to a brick court of the Dominican monastery in the town of Greifswald[25]. At the Praemonstratensian monastery of Belbuck/Białoboki near Treptow upon Rega/Trzebiatów the brickyard of the 13th/14th century[26] was excavated, an interesting kiln – perhaps connected with brickworks – was found in the settlement of the already mentioned Grobe Praemonstratensian monastery on the island of Usedom[27], and in the above mentioned monastery of Seehausen a brick

Figure 10. Liegnitz/Legnica, ducal residence: view from south-east (photo by C. Herrmann).

---

[23] Świechowski 2009, 530–548, with further literature.
[24] Świechowski 2009, 273–277, with further literature.
[25] Brandt et al. 2000; Ansorge 2005, 308–9.
[26] Biermann et al. 2007.
[27] Mangelsdorf et al. 2005.

kiln from the 15th century was located directly north of church and cloister[28].

Therefore it is no exaggeration to assign the Monastic construction workshops the crucial role of introducing this building material. Due to their wide-ranging communication network of builders and architects they had the possibility to import this technology to the northern and eastern territories which were lacking in stone material[29].

The origin of brick building

Traditionally it is assumed that the technique of burning bricks as a building material in the Middle Ages, starting in the middle and second half of the 12th century in several parts of middle and northern Europe, was a result of stimuli from north Italy, where Roman traditions were still vivid in the 12th century[30]. In the last years there has been growing support for the opinion that the rise of brick use occurred simultaneously at several places in Europe[31]. Some early brick churches at monasteries in the southern parts of the region between Elbe and Oder show quite strong stylistic ties to North Italy, for instance the apse of Doberlug (-Kirchhain, Lower Lusatia) with rows of small bows similar to San Michele in Cremona – both from the second half of the 12th century[32] – or famous Jerichow in the Altmark (after 1148) with the curious use of ceramic bowls as ornaments at the apse's façade (Figure 11). This is obviously an adoption of Northern Italian decoration forms[33].

To Pomerania the brick technology came apparently from Denmark, where brick was used already in the 1160/70s, at famous monastery churches like Esrum, Ringsted and Sorø on Zealand[34]. The Danish King Valdemar I the Great pursued a massive military expansion in the Slavic areas south of the Baltic, including Pomerania, from 1160 onward, and was able to subdue the princes of Rugia and the Greif Dukes to his dominion. Denmark had a strong political influence in this territory until 1227, when King Valdemar II the Victorious lost the battle of Bornhöved and his power in the Wend's Lands[35]. Only Rugia remained as a Danish fiefdom until 1325. With the Danish swords Christianity came to Rugia, and in the already more or less Christian countries – such as Pomerania – it accelerated Christian penetration. One effect of this are several early convents in Pomerania as filiae of Danish monasteries: Dargun (1172), Kolbatz (1173) and Eldena (1199) of Esrum in northern Zealand[36], Belbuck (1170-1180) of Lund (Drotten) in Scania, and in 1184 Grobe came under protection of Tommarp (Scania)[37]. Many of the early

Figure 11. Jerichow, Praemonstratensian Monastery church, ceramic bowl used as ornament in the apse, 12th century (photo by F. Biermann).

churches show architectonic-stylistic Danish influence, which is particularly visible in Eldena, Dargun, Kolbatz and Oliva[38]. In Eldena, allegedly the runic inscription of a Scandinavian name was found on a brick[39]. From this point of view the concentration of early Danish-style brick architecture on Rugia is not surprising – on the island the political and religious impact of the Danes was extremely intensive after 1168. Taken together, even the provenience of burned stone techniques in the coastal hinterland in the Southern Baltic[40] confirms the Danish influence.

Early brick tower castles in the hinterland of southern Baltic

Soon after 1200, several castles in the Pomeranian fiefdom were built of bricks, namely mighty towers. The most famous of them is the "Grützpott" at Stolpe upon Oder (today Brandenburg province): In a Slavic stronghold used since the 8th/9th century a roundish tower of 18m in diameter, 28m in height and up to 6m in thickness (Figure 12) was erected. The wall is constructed of bricks on erratic (granite) ashlars, but also of sandstone and limestone ashlars which perhaps were imported from Gotland and Scania. The interior of the building is arranged in quite a strange manner: Up to approximately 20 m height there is only one octagonal room, accessible only by the punched keystone of the vault made in fine workmanship. On the vault there exists a hall with fire place, toilet, window and the only entrance to the tower. It seems that further above there was originally one more floor planned. The tower is part of a motte castle, inserted as artificial mound in the northern part of the older stronghold. As a result of his very accurate-sophisticated research in Stolpe, J. C. Holst dates the mysterious building to the period 1202–1214, because

---

[28] Biermann et al. 2013b.
[29] See for instance Badstübner 1994, 34, 44.
[30] See for instance Badstübner 2009, 82.
[31] See for the discussion Perlich 2007.
[32] See Badstübner 2009, 79–81 Figures 4, 5.
[33] Perlich 2007.
[34] Krongaard Kristensen 2003, 95–98.
[35] Wille-Jørgensen 2003.
[36] Petersohn 1979, 447–9. Figure 8; Krongaard Kristensen 2003, 100–2.
[37] Flachenecker 2009, 327–9.; Olesen 2009.

[38] Krongaard Kristensen 2003, 100–2; Krongaard Kristensen 2009; Kratzke 2001; Reimann 2009.
[39] Möller 1998, 223.
[40] See for this beneath else Szczesiak 2005, 402; Biermann 2009, 27.

Figure 12. Stolpe upon Oder, the castle tower called "Grützpott", 13th century (photo by F. Biermann).

of its construction technique, its architecture and historical contemplations[41].

Another giant brick tower on ashlars was erected in the important castle of Haus Demmin in Cispomerania. The ruined preserved building was rectangular on its ground floor and had a side length of 15–16m. At the level of the basement the walls are convex. The interior of the tower, apparently also the exterior of its upper parts, was round. The tower, also part of a motte castle in the area of an older Slavic stronghold, is linked by the chronicler Detmar of Lübeck of the late 14th century to the fortification of Haus Demmin by the Danes in the year 1211[42]. Another tower of this kind, a round one of 19m diameter and 6m thick walls, was identified in the Alt Kalen stronghold (Cispomerania) in the 19th century, but it is not documented in detail. It is not known whether it was made of bricks or not[43].

These towers are often connected with Danish influence, even interpreted as Danish tower castles. They are supposed to have served as military and administrative centers of Danish sovereignty south of the Baltic, in particular also as symbols of Danish power. Indeed these towers were built abruptly in the time of Danish dominion in Pomerania and its neighboring regions, and the Danes used bricks already frequently in this period – not only in the sacral but also in the profane context. To give only a few examples: King Valdemar I rebuilt the Danewerk border fortification in Schleswig by using bricks (approximately in 1170), which is mentioned with pride on the lead plate in the King's grave in Ringsted, dating from 1182[44]. His castle on the small island of Sprogø in Storebælt, also mentioned on the plate,

Figure 13. Salzwedel, castle tower, late 12th/13th century (photo by F. Biermann).

was according to archaeological research constructed partly of bricks as well, but not in the shape of a large tower[45]. The chronicler Detmar of Lübeck – writing in the 14th century – reports for the year 1217 that Count Albert, by order of Valdemar II, built a large brick tower in Travemünde on the Mecklenburg coast[46]; hence Denmark is apparently a possible point of origin.

However – the problem of the Danish hypothesis is that there are no examples of large brick-built round towers in Denmark or Scania from the 12th or early 13th century. The measures of the best-researched Danish Tower castle of the 12th century, the round Bastrup Tower on Zealand – 21m in diameter, wall thickness of more than 6m – fits to the southern examples but was constructed without brick: The walls of the impressive building are made of calcareous tufa[47]. Expected, in few cases also proven, are rectangular towers in 12th/early 13th century royal castles in Denmark. Round ones are not known, and according to the current state of research no brick towers were constructed either[48].

---

[41] Schulz 1998; Möller 1998, 221–2.; most important: Holst 2009.
[42] Möller 1998, 219–21, Figures 1, 2; Holst 2009, 98 footnote 14; Jänicke et al. 2009. The chronicler brother Detmar wrote in his annals: „In deme jare Christi 1211 do wan koningh woldemer den hertoghen af van Stettin vele eres landes, und buwede demyn wedder". Chronik 1829, 86; for motte castles in Cispomerania see Möller 1993.
[43] Möller 1998, 223; Holst 2009, 98.
[44] See Etting 2010, 17.
[45] The small castle consists of a rectangular curtain wall made of a granite boulder of 26 to 32m side length, some buildings and perhaps a tower, but there is no clear evidence for it; Engberg 2009.
[46] "In demesulven jare christi [1217] do buwede desulve greve Albert van des koningh Woldemares weghene Travenemunde, unde leghde dar enen vasten torn von teghele" (Chronik 1829, 93).
[47] See Etting 2010, 16, with further literature.
[48] See Wille-Jørgensen 2001, 169–71 Figure 5.

While we do not know large brick towers from the period around 1200 in Denmark, several such buildings are known from the territory south of the Baltic. For instance the Askanian Margraves of Brandenburg erected mighty brick towers in Salzwedel – with 18 m in diameter, dendrochronologically dated to the 1190s (Figure 13)[49] – and in (Berlin-) Spandau, the so-called "Juliusturm" attributed to Margrave Albrecht (1205–1220)[50]. Regarding these buildings there is no reason to assume Danish influence. They indicate a rather dissenting interpretation: Apparently several important magnates in the eastern Elbe region started building such towers as instruments and symbols of power in the decades round and shortly after 1200, in co-orientation to each other the Greif dukes as well as the Askanians, smaller rulers and – referring to Travemünde – also larger ones than those built by the Danish Kings[51]. Thus the sole Danish deduction of the towers in the south of the Baltic requires further discussion.

Conclusion

Many questions referring to the beginnings and the origin of brick building in Pomerania, Pomerelia and Lower Silesia are still unanswered, but art historical, building historical and archaeological research have delivered important information in the last decades. Brick architecture in these regions started between 1180 and 1195, but it did not increase before the 13th century and never completely replaced ashlars and fieldstone. The provenance of the new techniques is not absolutely clear, but several arguments underline the hypothesis of important Danish influence in the north. In any case the significance of the monasteries for the brick technology cannot be overestimated. They played a leading role in the innovation, the dissemination and the economic development of brick production and use.

References

Ansorge, J. 2005. Kalkbrennerei und Ziegelherstellung. In H. Jöns, F. Lüth and H. Schäfer (eds), Archäologie unter dem Straßenpflaster. 15 Jahre Stadtkernarchäologie in Mecklenburg-Vorpommern. Beiträge zur Ur- und Frühgeschichte Mecklenburg-Vorpommerns 39, 307–310. Schwerin, Archäologisches Landesmuseum

Appelt, H. 1971. Die mittelalterliche deutsche Siedlung in Schlesien. In Deutsche Ostsiedlung in Mittelalter und Neuzeit, 1–19. Köln-Wien, Böhlau

Badstübner, E. 1994. Feldstein und Backstein als Baumaterial in der Mark Brandenburg während des 12. und 13. Jahrhunderts. architectura, 34-45

Badstübner, E. 2009. Grundlagen und Ausprägungen der Backsteingotik im Ostseeraum. In O. Auge, F. Biermann and C. Herrmann (eds), Glaube, Macht und Pracht. Geistliche Gemeinschaften des Ostseeraums im Zeitalter der Backsteingotik. Archäologie und Geschichte im Ostseeraum 6, 77–94. Rahden/Westfalen, Leidorf

Benl, R. 1999. Pommern bis zur Teilung von 1368/72. In W. Buchholz (ed.), Deutsche Geschichte im Osten Europas – Pommern. Berlin, Siedler, 21–126

Biermann, F. 2006. Die Kirchen des Bischofs Otto von Bamberg in Pommern – ein Beitrag zur Frühgeschichte der Kirche St. Paul in Usedom und zur Lage der missionszeitlichen Sakralbauten im Odermündungsgebiet. In F. Biermann, M. Schneider and Th. Terberger (eds), Pfarrkirchen in den Städten des Hanseraums. Archäologie und Geschichte im Ostseeraum 1, Rahden/Westfalen, Marie Leidorf, 21–38

Biermann, F. 2009. Einführung. In O. Auge, F. Biermann and C. Herrmann (eds), Glaube, Macht und Pracht. Geistliche Gemeinschaften des Ostseeraums im Zeitalter der Backsteingotik. Archäologie und Geschichte im Ostseeraum 6, 9–37. Rahden/Westfalen, Leidorf

Biermann, F. 2012. Der mittelalterliche Nonnenkonvent auf dem Klosterberg von Altentreptow. Bodendenkmalpflege in Mecklenburg-Vorpommern, Jahrbuch 2011-59, 107–160

Biermann, F., Blum, O. and Hergheligiu, C. 2013a. Neue Forschungen zum Prämonstratenserstift Grobe auf Usedom – Ausgrabungen am Wilhelmshofer Priesterkamp im Jahre 2010. Bodendenkmalpflege in Mecklenburg-Vorpommern 2013, in print

Biermann, F., Frey, K. and Meyer, C. 2013b. Erste Einsichten zur Baugestalt des mittelalterlichen Zisterzienserinnenklosters Seehausen in der Uckermark. In C. Theune, G. Scharrer-Liška, E. H. Huber and Th. Kühtreiber (eds.), Stadt – Land – Burg. Festschrift für Sabine Felgenhauer-Schmiedt zum 70. Geburtstag. Internationale Archäologie, Studia honoraria 34, 295–308. Rahden/Westfalen, Leidorf

Biermann, F. and Rębkowski, M. 2007. Usedom-Grobe und Treptow (Trzebiatów)-Belbuck – Herrschafts- und Sakraltopographie pommerscher Zentralorte im 12./13. Jahrhundert. In G. H. Jeute, J. Schneeweiß and C. Theune (eds), aedificatio terrae. Beiträge zur Umwelt- und Siedlungsarchäologie Mitteleuropas. Festschrift für Eike Gringmuth-Dallmer. Studia honoraria 26, 69–78. Rahden/Westfalen, Leidorf

Brachmann, H., Foster, E., Kratzke and C./Reimann, H. 2003. Das Zisterzienserkloster Dargun im Stammesgebiet der Zirzipanen. Ein interdisziplinärer Beitrag zur Erforschung mittelalterlicher Siedlungsprozesse in der Germania Slavica. Stuttgart, Franz Steiner

Brandt, D. and Lutze, A. 2000. Ein mittelalterlicher Kalk-/Ziegelhof am Ryck bei Greifswald. In U. Müller (ed.), Handwerk – Stadt – Hanse. Ergebnisse der Archäologie zum mittelalterlichen Handwerk im südlichen Ostseeraum. Greifswalder Mitteilungen 4, 145–160. Frankfurt, Peter Lang

Büttner, B. 2007. Die Pfarreien der Insel Rügen. Von der Christianisierung bis zur Reformation. Veröffentlichungen der Historischen Kommission für Pommern 42, Köln, Böhlau

Chronik des Franziskaner Lesemeisters Detmar, ed. by F. H. Grautoff, 1. Part. Die lübeckischen Chroniken in niederdeutscher Sprache 1. Hamburg: Perthes (1829)

---

[49] Kind personal communication by U. Frommhagen, Salzwedel.
[50] Gehrke 1991, 118–120.
[51] See Möller 1998, 223.

Engberg, N. 2009. The castle on Sprogø – a royal castle from 12th century. In B. Fløe Jensen and D. Wille-Jørgensen (ed.), Expansion – Integration? Danish-Baltic Contacts 1147–1410 AD, 67–74. Vordingborg, Danmarks Borgcenter

Etting, V. 2010. The royal castles of Denmark during the 14th century. An analysis of the mayor royal castles with special regard to their functions and strategic importance. Publications of the National Museum, Studies in Archaeology and History 19, Copenhagen, National Museum of Denmark

Fait, J. 1957/58. Die Ruine in Stolpe an der Peene und ihre Bedeutung für die romanische Baukunst in Nordostdeutschland. Wissenschaftliche Zeitschrift der Ernst-Moritz-Arndt-Universität Greifswald, Gesellschafts- und sprachwissenschaftliche Reihe 8, 1957/58, 203–217

Fait, J. 1963. Die Benediktinerkirche in Stolpe an der Peene. Ein Ausgrabungsbericht und Rekonstruktionsversuch. Greifswald-Stralsunder Jahrbuch 3, 119–134

Feldmann, H.-C. 2000. Mecklenburg-Vorpommern. Georg Dehio Handbuch der Deutschen Kunstdenkmäler. München-Berlin, Deutscher Kunstverlag

Flachenecker, H. 2009. Die Rolle der Prämonstratenser im Ostseeraum. In O. Auge, F. Biermann and C. Herrmann (eds), Glaube, Macht und Pracht. Geistliche Gemeinschaften des Ostseeraums im Zeitalter der Backsteingotik. Archäologie und Geschichte im Ostseeraum 5, 323–338. Rahden/Westfalen, Leidorf

Gehrke, W. 1991. Das Gelände der Spandauer Zitadelle im Mittelalter. In Berlin und Umgebung. Führer zu archäologischen Denkmälern in Deutschland 23, 117–124. Stuttgart, Theiß

Grodecki, R. 1933. Dzieje polityczne Śląska do roku 1290. In Historia Śląska od najdawniejszych czasów do roku 1400 1, 155–326. Kraków, Academy

Grünhagen, C. 1884. Geschichte Schlesiens I. Bis zum Eintritt der habsburgischen Herrschaft 1527. Gotha, Perthes

Hardt, M. 2005. Die Veränderung der Kulturlandschaft in der hochmittelalterlichen Germania Slavica – offene Fragen beim derzeitigen Forschungsstand. In F. Biermann and G. Mangelsdorf (eds), Die bäuerliche Ostsiedlung des Mittelalters in Nordostdeutschland. Untersuchungen zum Landesausbau des 12. bis 14. Jahrhunderts im ländlichen Raum. Greifswalder Mitteilungen 7. Frankfurt/Main, Lang, 17–28

Heußner, K. U. 2005. Handel mit Holz. In H. Jöns, F. Lüth and H. Schäfer (eds), Archäologie unter dem Straßenpflaster. 15 Jahre Stadtarchäologie in Mecklenburg-Vorpommern. Beiträge zur Ur- und Frühgeschichte Mecklenburg-Vorpommerns 39. Schwerin, Archäologisches Landesmuseum, 125–128

Holst, J. C. 2005. Neues zur Baugeschichte von St. Marien. In Der Klosterhof und die Kirche St. Marien in Bergen auf Rügen. Bergen: Stadtverwaltung, 22–48

Holst, J. C. 2009. The Tower named "Grüttpott" at Stolpe upon Oder. In B. Fløe Jensen and D. Wille-Jørgensen (ed.), Expansion – Integration? Danish-Baltic Contacts 1147–1410 AD, 95–118. Vordingborg, Danmarks Borgcenter

Jänicke, R. and Schanz, E. 2009. Von Slawen und Deutschen – Die lange Kontinuität der Burganlage „Haus Demmin", Lkr. Demmin. In Archäologische Entdeckungen in Mecklenburg-Vorpommern. Kulturlandschaft zwischen Recknitz und Oderhaff. Archäologie in Mecklenburg-Vorpommern 5, 185–186. Schwerin, Landesamt für Kultur und Denkmalpflege

Kloer, H. 1929. Kloster Eldena. Kunstwissenschaftliche Studien I. Berlin, Deutscher Kunstverlag

Kratzke, C. 2001. Die Architektur der Zisterzienser im Ostseegebiet: Gemeinsamkeiten, Unterschiede, Einflüsse. In O. Harck and C. Lübke (eds), Zwischen Reric und Bornhöved. Die Beziehungen zwischen den Dänen und ihren slawischen Nachbarn vom 9. bis ins 13. Jahrhundert. Beiträge einer internationalen Konferenz, Leipzig 4.-6. Dezember 1997. Forschungen zur Geschichte und Kultur des östlichen Mitteleuropa 11, 197–225. Stuttgart, Franz Steiner

Krongaard Kristensen, H. 2003. Backsteinarchitektur in Dänemark bis zur Mitte des 13. Jahrhunderts. In M. Gläser, D. Mührenberg and P. Birk Hansen (eds), Dänen in Lübeck 1203 – 2003. Danskere i Lübeck. Ausstellungen zur Archäologie in Lübeck 6, 94–103. Lübeck, Schmidt-Römhild

Krongaard Kristensen, H. 2009. Architectural relations between Danish Cistercian churches and the Daughters of Esrum at Dargun, Eldena and Kolbatz. In O. Auge, F. Biermann and C. Herrmann (eds), Glaube, Macht und Pracht. Geistliche Gemeinschaften des Ostseeraums im Zeitalter der Backsteingotik. Archäologie und Geschichte im Ostseeraum 5, 59 –75. Rahden/Westfalen, Leidorf

Mangelsdorf, G. 1999. Neue Ausgrabungen in der Klosterruine von Eldena bei Greifswald. In G. Mangelsdorf (ed.), Von der Steinzeit zum Mittelalter, 225–291. Greifswalder Mitteilungen 3. Frankfurt/M., Peter Lang

Mangelsdorf, G., N. Benecke and F. Biermann 2005. Untersuchungen zum frühgeschichtlichen Wirtschafts- und Herrschaftszentrum Usedom II: Die spätslawische Siedlung am Priesterkamp. Bodendenkmalpflege. In Mecklenburg-Vorpommern, Jahrbuch 2004-52, 397–545

Möller, G. 1993. Adlige Befestigungen in Vorpommern vom Ende des 12. - Anfang des 17. Jh.s. In Castella Maris Baltici 1, 149–153 Stockholm.

Möller, G. 1998. Die Anfänge "deutschen" Burgenbaus in Vorpommern. Ethnographisch-Archäologische Zeitschrift 39, 1998, 217–228

Moździoch, S. 1990. Organizacja gospodarcza państwa wczesnopiastowskiego na Śląsku. Studium archeologiczne. Wrocław-Warszawa-Kraków, Ossolineum

Nawrocki, P. 2010. Der frühe dänische Backsteinbau, Studien zur Backsteinarchitektur 9. Berlin, Lukas Verlag

Olesen, J. E. 2009. Der Einfluss dänischer Klöster auf den Ostseeraum. In O. Auge, F. Biermann and C. Herrmann (eds), Glaube, Macht und Pracht. Geistliche Gemeinschaften des Ostseeraums im Zeitalter der

Backsteingotik. Archäologie und Geschichte im Ostseeraum 5, 49–57. Rahden/Westfalen, Leidorf

Perlich, B. 2007. Mittelalterlicher Backsteinbau in Europa. Zur Frage nach der Herkunft der Backsteintechnik. Berliner Beiträge zur Bauforschung und Denkmalpflege 5. Petersberg, Imhof

Petersohn, J. 1979. Der südliche Ostseeraum im kirchlich-politischen Kräftespiel des Reichs, Polens und Dänemarks vom 10. bis 13. Jahrhundert. Mission – Kirchenorganisation – Kultpolitik. Ostmitteleuropa in Vergangenheit und Gegenwart, 17. Köln-Wien, Böhlau

Quandt, L. 1856. Ostpommern, seine Fürsten, fürstlichen Landestheilungen und Districte. Baltische Studien 16, 197–156

Rębkowski, M. 2011. Die Christianisierung Pommerns. Eine archäologische Studie. Universitätsforschungen zur prähistorischen Archäologie 197. Bonn: Habelt

Reimann, H. 2009. Dänische Einflüsse auf Zisterzienserklöster im slawischen Siedlungsgebiet. Das Beispiel Dargun. In O. Harck and C. Lübke (eds), Zwischen Reric und Bornhöved. Die Beziehungen zwischen den Dänen und ihren slawischen Nachbarn vom 9. bis ins 13. Jahrhundert. Beiträge einer internationalen Konferenz, Leipzig 4.-6. Dezember 1997. Forschungen zur Geschichte und Kultur des östlichen Mitteleuropa 11, 187–195. Stuttgart, Franz Steiner

Reimann, H., Ruchhöft, F. and Willich, C. 2011. Rügen im Mittelalter. Eine interdisziplinäre Studie zur mittelalterlichen Besiedlung auf Rügen. Forschungen zur Geschichte und Kultur des östlichen Mitteleuropa 36. Stuttgart, Franz Steiner

Ruchhöft, F. 2010. Die Burg am Kap Arkona. Götter, Macht und Mythos. Archäologie in Mecklenburg-Vorpommern 7, Schwerin, Landesamt für Kultur und Denkmalpflege, 61–67

Schäfer, H. 2005. Archäologische Untersuchungen im ehemaligen Nonnenkloster zu Bergen. In Der Klosterhof und die Kirche St. Marien in Bergen auf Rügen, 17–21. Bergen, Stadtverwaltung

Schulz, R. 1998. Stolpe, eine Turmburg des späten 12. Jahrhunderts an der Oder. Eine Befestigung der Dänen in Pommern gegen die Markgrafen von Brandenburg? Château Gaillard 18, 211–221

Seip Bartholin, Th. 1995. Dendrochronologisk undersøgelse af kirken og klokkehuset I Altenkirchen, Rygen, Tyskland. Nationalmuseets Naturvidenskabelinge Undersøgelser, rapport 5. København, Nationalmuseet

Świechowski, Z. 2009. Katalog architektury romańskiej w Polsce. Warszawa, Wydawnictwo DiG

Szczesiak, R. 2005. Stein gewordene Zeugnisse des Glaubens – Architektur und Kunst der Klöster und Stifte. In H. Jöns, F. Lüth and H. Schäfer (eds), Archäologie unter dem Straßenpflaster. 15 Jahre Stadtkernarchäologie in Mecklenburg-Vorpommern. Beiträge zur Ur- und Frühgeschichte Mecklenburg-Vorpommerns 39, 401–410. Schwerin, Archäologisches Landesmuseum

Tintelnot, H. 1951. Die mittelalterliche Baukunst Schlesiens. Kitzingen, Holzner-Verlag

Wille-Jørgensen, D. 2001. Die Burg Vordingborg als Basis dänischer Eroberungszüge an die slawische Ostseeküste. In O. Hardt and C. Lübke (eds), Zwischen Reric und Bornhöved. Die Beziehungen zwischen den Dänen und ihren slawischen Nachbarn vom 9. bis ins 13. Jahrhundert. Beiträge einer internationalen Konferenz, Leipzig 4.-6. Dezember 1997. Forschungen zur Geschichte und Kultur des östlichen Mitteleuropa 11, 165–177. Stuttgart, Franz Steiner

Wille-Jørgensen, D. 2003. Das Ostseeimperium der Waldemaren. Dänische Expansion 1160–1227. In M. Gläser, D. Mührenberg and P. Birk Hansen (eds), Dänen in Lübeck 1203 – 2003. Danskere i Lübeck. Ausstellungen zur Archäologie in Lübeck 6, 26–35. Lübeck, Schmidt-Römhild

Zaske, N. 1969. Ausgrabungs- und Forschungsergebnisse zum norddeutschen Backsteinbau des Mittelalters. Wissenschaftliche Zeitschrift der Ernst-Moritz-Arndt-Universität Greifswald, Gesellschafts- und sprachwissenschaftliche Reihe Nr.3, 4, Jg. 18, 371–379

Brick in Historic Architecture:
The Research Experience of Warsaw

Włodzimierz Pela

Abstract: The pace and scope of the documentation work conducted since 1945 in the course of removing the rubble remains of the war-damaged historic buildings of Warsaw's Old and New Town necessitated the establishment and application of appropriate methods of measurement and research. For this purpose, a special recording card for brick walls was developed. This included measurements, notes and sketches, descriptions of state of preservation as well as measurements made of undisturbed brickwork courses. In this way, documentation allowing differentiation of the original fabric was made, which was important for designing the conservation programme. Analysis of the results and comparative material produced much new information on medieval and post-Medieval brickwork. One result was the creation in the 1950s of a table of Warsaw brick sizes from the 14th to 19th centuries, this could be used to differentiate different chronological stages of a brick construction. This is despite the objections from historians that written sources do not confirm the standardization of Medieval brick sizes. Later studies have modified some of the chronological details of this scheme, but it is undisputed that the size of bricks and its dimensions changed during this period.

Keywords: Archaeological and architectural examinations, bricks, methods of measurement and research, Poland, Warsaw

Introduction

In research on the history of architecture, an interest in brick as a building material dates back at least to the nineteenth century. The topic has been studied by many investigators of architecture, expressing their observations both in publications describing historic buildings, as well as in synthetic publications. The subjects of interest were the many issues related to brick architecture: the origins of the use of brick in construction in various parts of Europe, its technology and manufacturing techniques, the variety of ways bricks were laid in a wall. Attempts were also made to make use of the changing size of bricks to date the constructions in which they were used. The names of the many researchers involved in such studies include Friedrich Wachtsmuth, Konrad Werner Schulze, Otto Stiehl, Friedrich Fischer, Nathaniel Lloyd.[1] The names of other researchers, due to the fact that they published their work in more exotic languages like Polish, are however much less known and I would like in my presentation to discuss the achievements of some of them, working in twentieth-century Warsaw.

Brick dimension analyses for dating masonry in the Old Town (1913)

One of the first examples of architectural analysis of the historical buildings in Warsaw which includes measurement of the size and arrangement of the bond brick building is the history of the development of Old Town Square 31 Przyrynkowa Street and 48 Piwna Street. Here, architectonic (called archaeological) research was carried out during restoration work by architect Jarosław Wojciechowski in 1913. The results were published several years later in 1928, in a monographic study of the building.[2]

On the basis of the measurements of the bricks, Jarosław Wojciechowski defined the chronological position of various parts of the building belonging to different historical styles. In his fieldwork he differentiated several types of brick:

- Gothic brick, changing in size in the period from the fourteenth to the early sixteenth century (Earlier Medieval, 9–10 x 14 x 28cm, Later measuring 8–8 ½ x 12 x 27cm),
- Renaissance brick of the sixteenth century (7 x 13 x 26cm), and
- Bricks of Italian type of the seventeenth century (after 1630 – 6 ½ x 13 ½ x 25cm).

The study took into account the observations of other researchers, among others, the German conservator Julius Kohte working in the Poznan region in the nineteenth century,[3] stressing, however, that the millimetre accuracy of the German school was not required.

Developing the methodology of brick dimension analysis during the rebuilding of the Old Town after 1945

Another important step in the research on the brick architecture of Warsaw took place in the course of removing rubble and rebuilding of the historic structures in the city's historic core, the Old and New Town, which had been

---

[1] Stiehl 1898, 1923, 1927, Wachtsmuth 1925, 1938; Lloyd 1925, Schulze 1927, Fischer 1944.
[2] Baruch and Wojciechowski 1928.
[3] Kohte 1895–1898.

Figure 1. Aerial photograph of the destroyed Old Town, 1945.

burnt down and blown up in 1944 (Figure 1). The scientific community was aware that in many cases this was the last chance to document the remains of the old buildings of the city. The pace and scope of construction work started in 1945 forced those directing this documentation to identify and apply effective research methods of measurement (Figure 2). These had to produce relatively satisfactory results, even though these were to be achieved in a short time with the employment of many people (e.g. students) who were not always fully appraised of the nature of the research material (Figure 3 and 4).

For this purpose, the staff of the Department of Polish Architecture of the Technical University of Warsaw and especially the young architect directing the work, Zdzisław Tomaszewski, created a special record card for use in documenting the surface of the walls. It contained, among other things measured notes and sketches, descriptions of the state of preservation, and above all measurements of a minimum of 20 of the bricks made on undisturbed segments of the brick wall. To standardize the description, two types of brick recording forms were introduced. The first (Figure 5), more elaborate, was used in areas of particular importance because of the chronology. The second form (Figure 6) was used primarily to gather statistical comparative material.[4]

In this manner, in the years 1949–1956, a rich database of information describing the actual state of the remains of the old buildings was created. This allowed the differentiation

Figure 2. Documentation of the walls beside number 12 Szeroki Dunaj Street. Architectural investigation in 1952.

---

[4] Tomaszewski 1956, 41–46.

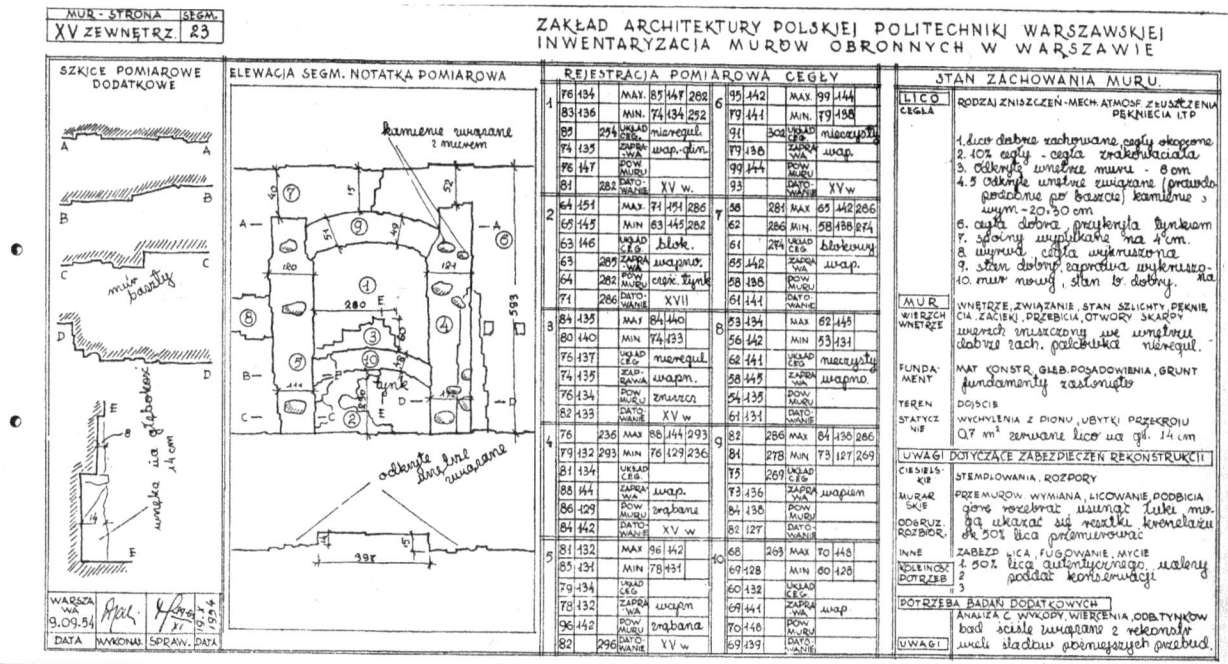

Figure 3. Department of Architecture, the Warsaw Technical University. Drawing documentation from the architectural investigation in 1952.

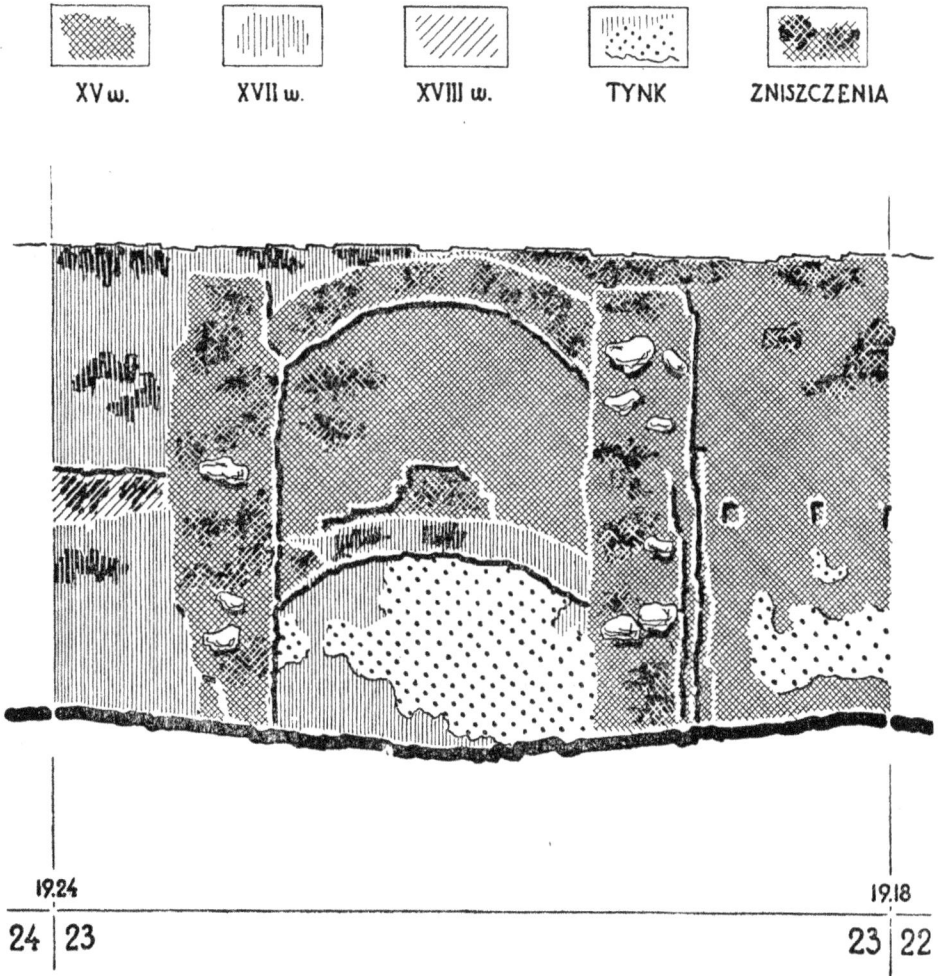

Figure 4. Department of Architecture, the Warsaw Technical University. Recording form of the defence walls of Old Town. Documentation from the architectural investigation in 1952.

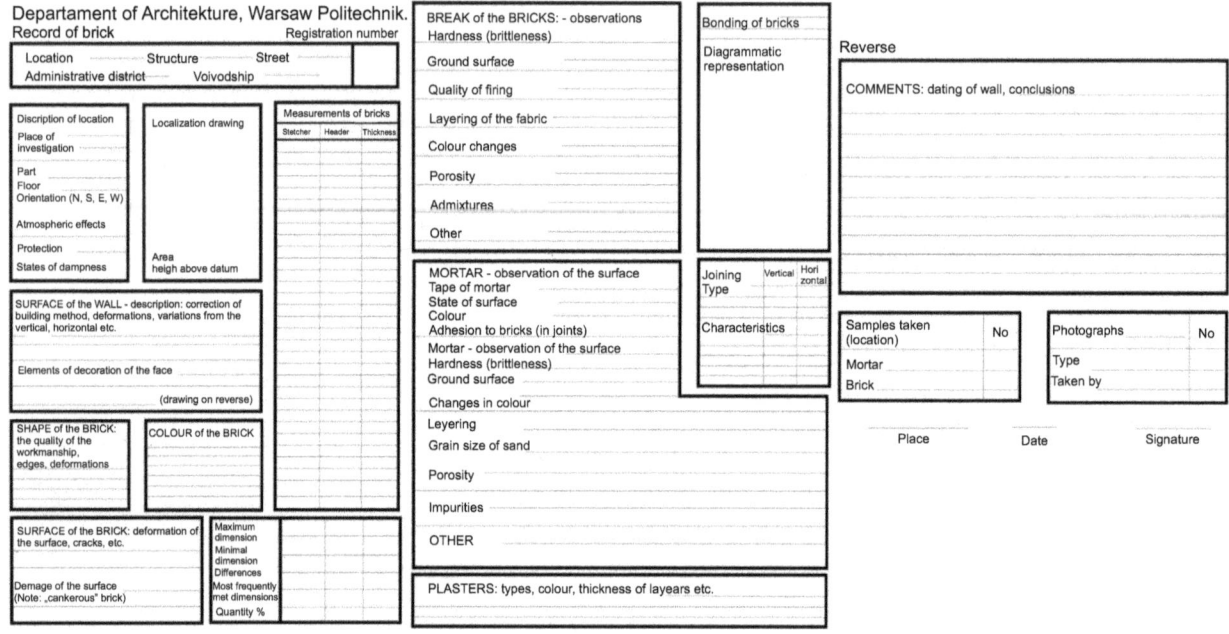

Figure 5. Department of Architecture, the Warsaw Technical University. The architectural investigations. The brick recording form – more elaborate, was used in areas of particular importance for the chronology.

Figure 6. Department of Architecture, the Warsaw Technical University. The brick recording form – used primarily to gather statistical comparative material.

of the authentic pieces of masonry material in these structures for purposes of research, conservation projects and reconstruction work.

On the basis of the measurement of materials collected during fieldwork conducted at the Castle in Warsaw and in the Old Town area, further analyses and comparisons were conducted, comparing material of the Middle Ages and later date. One of the results of the work was the creation and publication of a table of the sizes of bricks used in construction in Warsaw from the fourteenth to the nineteenth centuries.[5]

This creation of this table utilised simple statistical methods to analyse a number of factors that would affect the accuracy of measurements, among others, changes occurring in the production process (shrinkage) and also studying the results of the physico-chemical measurements. Alongside the publication of Zdzisław Tomaszewski, research by Hanna Jabłczyńska-Jędrzejewska was conducted on the mortars used in the medieval building of Warsaw.[6]

The table of dimensions of bricks used in the construction of walls in Warsaw from, the 14th to the 20th centuries

The results of the study allowed the description of the variations in brick thickness during the fourteenth-twentieth centuries (Figure 7). At the beginning of the fourteenth century, the dimensions were approximately 10cm, and

---

[5] Tomaszewski 1955, 31–52.
[6] Jabłczyńska-Jędrzejewska 1958, 85–94.

Date	Dimensions in millimetres		
	thickness	header	stretcher
XIV (the first half)	90 - 102	120 - 130	260 - 280
XIV (the second half)	100 - 115	125 - 140	280 - 300
XIV/XV	84 - 95	115 - 135	265 - 285
XV	76 - 87	120 - 130	255 - 280
XV/XVI	72 - 85	120 - 140	260 - 275
XVI (the second half)	67 - 82	125 - 140	255 - 280
XVII (the first half)	62 - 74	120 - 140	265 - 290
XVII (the second half)	56 - 68	125 - 135	255 - 285
XVIII (the first half)	54 - 62	125 - 140	255 - 280
XVIII (the second half)	48 - 56	130 - 145	260 - 285
XIX	65	130	270
XX (the first half)	60	130	270
XX (the second half)	65	120	250

Figure 7. Table of the sizes of Warsaw bricks (by Z. Tomaszewski, 1955).

reached a maximum of 11.5cm in the second half of the fourteenth century. The thickness gradually declines through the period from the fifteenth to seventeenth centuries. By the second half of the seventeenth century, the thickness is a minimum of about 5.5cm. The breadth of the brick varies between 12 and 14cm, being the narrowest in the fourteenth and fifteenth centuries and broadest in the eighteenth century. The length had a large deviation from 25.5 to 29cm; the changes are very irregular, therefore making their suitability for dating less important. The sequence of these changes could be given absolute dates by the measurement of bricks in structures which could be dated in the written sources.

The results of this research were used to investigate the history of the various buildings in Warsaw, as well as for the entire built-up area of the Old Town. For example, it was possible to differentiate the various phases of the expansion of the building at 21 Old Town Square, where the dimensions of the bricks allowed the recognition within the existing remains of the original fourteenth to fifteenth century two-range building. In the latter, the partition wall between the ranges was rebuilt and vaulting was added at the turn of the fifteenth and sixteenth centuries, and then at the end of the sixteenth century, a third range was added (Figure 8).

An example of the use of the method for the entire area of the Old Town would be the execution and publication of a plan in 1954, which covered the cellars as well presented location and vaulting of Gothic walls in the urban complex. In this plan, the Gothic walls were shown at the ground floor level, as well as at the level of the first and second floors (Figure 9).

Measurement and analysis of the brickwork was also used to develop an inventory for the conservation of the defensive walls surrounding the Old City, showing the intact original surface and defining areas of later rebuilding and repair.

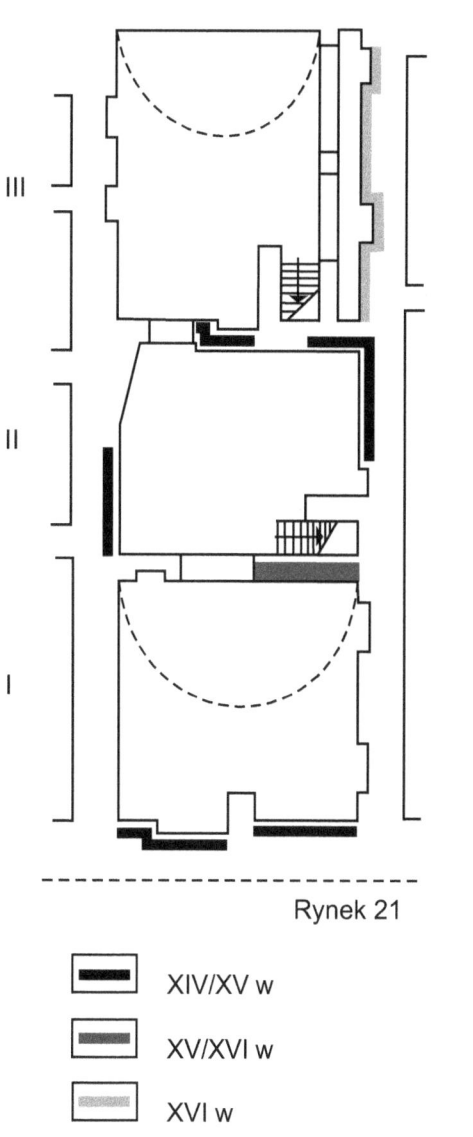

Figure 8. Old Town Square no. 21. Determination of the phasing of rebuilding of some structures in the Old Town.

Figure 9. Warsaw Old Town, remains of Gothic building. An example of the use of the entire area of the Old Town would be the execution and publication of the plan in 1954 of the cellars with the presentation of the location of Gothic walls and vaulting.

Unfortunately, in 1956, Tomaszewski – the initiator and enthusiastic propagator of the use of these methods of research on brick architecture – left Poland and did not complete his planned historical research covering the use of ceramic building materials in the Warsaw region (the historical district of Mazovia) during the period from the twelfth to the nineteenth centuries.

Modification of the methodology of brick dimension analysis – the use of a graphic scheme (1960)

A few years later, in 1957–1963, the town's defensive walls were again studied by the Department of Polish Architecture of the Warsaw Technical University. This was carried out under the direction of Jarosław Widawski during the conservation of the segment of walls from the Royal Castle to the Marszałkowska Tower. Like his predecessors, he devoted quite a bit of attention to the study of the bricks, and utilised an analysis of their dimensions in order to determine the relative chronology of the individual parts of the investigated masonry. In his opinion, the comparison of the dimensions of the bricks used in the construction was the final proof of the correctness of the phasing determined in the earlier stages of the work of the defensive walls of the Old Town, and in particular the older defensive wall. This research enabled him to determine the history of the construction of the walls.[7]

The method developed by Zdzisław Tomaszewski was put into graphic form in the research of Andrzej Gruszecki in the 1960s.[8] While conducting architectonic research in the royal residence at Ujazdów Castle, just outside the town, he came to the conclusion that the use of hundreds of record cards would be quite a complex task without the use of electronic computers. Therefore he replaced digital values by a graphic means of showing the brick dimensions, allowing the free and easy comparison of thousands of measurements. He considered that the graphic means of showing the dimensions would allow different size-groups of bricks to be defined among the demolition material used on the site, which generally had a wide distribution. This would be indicated by clusters of different spots on the plan. The use of this method, in his opinion, allowed a more legible differentiation of the groups of bricks used for the

[7] Widawski 1970, 239–252.
[8] Gruszecki 1965, 55–58.

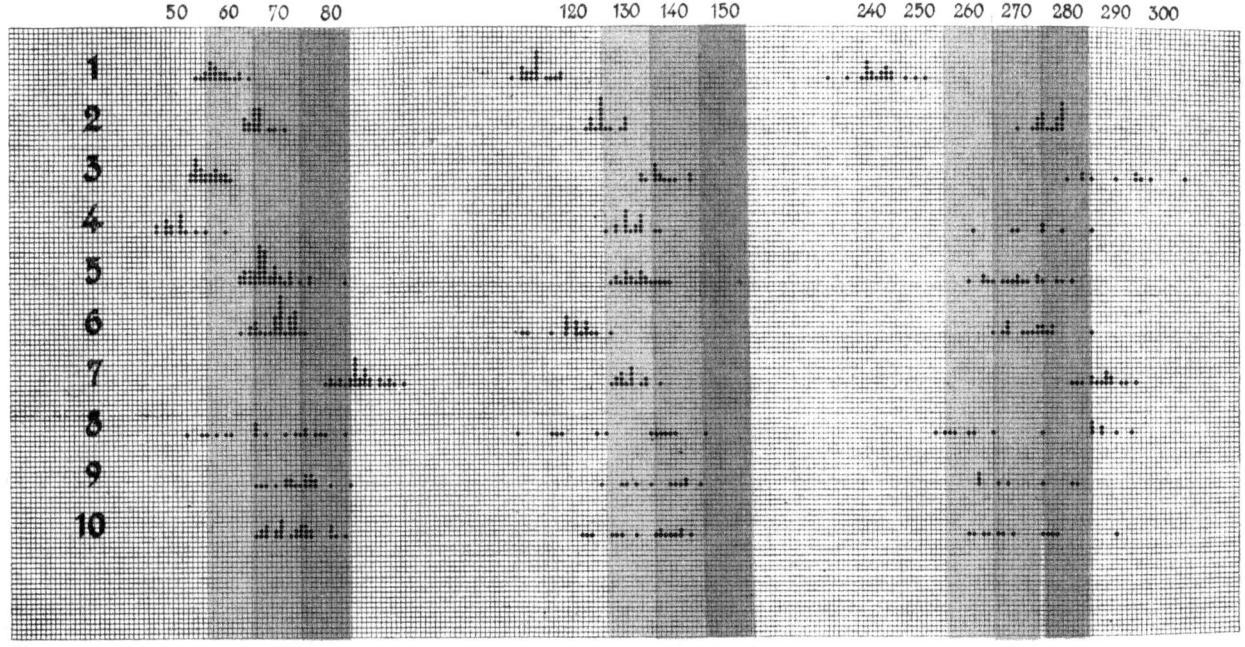

Figure 10. Example of the use of graphic means of displaying brick dimensions – created by A. Gruszecki.
1 – modern brick of 1962,
2 – brick from the 19th century Russian rebuilding of the Ujazdowski Castle in Warsaw,
3 – brick from the 1766–1772 rebuilding of the Ujazdowski Castle,
4 – brick from the rebuilding of the Ujazdowski Castle in the 1730s,
5 – brick from the seminary building in Tykocin (1630–1640),
6 – brick from the church of St John in Wizna consecrated in 1500,
7 – brick from the second phase of the rebuilding (1559) of the church of St Mary in Chojna,
8 – brick from the demolition of the Ujazdowski Castle,
9 and 10 – two measurements of bricks taken independently of each other from the same wall in Tykocin.

construction of a single architectural structure. In the case of the Ujazdowski Castle, it also showed the differences in the types of bricks used for the construction of the same part of the wall, which was the result of using products of several brickyards together. Gruszecki's publication is illustrated by a graphic summary, which shows the measurements of different batches of bricks from the walls of Ujazdowski Castle and other Polish buildings, such as the seminary in Tykocin, the church of John the Baptist in Wizna and the Church of St. Mary in Chojna (Figure 10).

Viewpoint of historians on the use of brick dimension analysis in the study of historic architecture

This method of measurement used to determine the chronology of brick architecture encountered opposition from a number of historians who had found no written confirmation of standardization in the production of bricks in the Middle Ages.[9] The authors of these studies also expressed caution due to the pioneering nature of the research. They believed that the most secure method was to compare the dimensions of the bricks within a single building, but care should be taken in making wider generalizations, because many of the processes involved in the production of bricks were different in various periods and regions. In addition, the changes in the dimensions of the bricks are more visible over longer periods of time, e.g., half a century, but the period of a quarter-century is still difficult to grasp by this method. On the basis of results obtained in later fieldwork, it also seems necessary to make adjustments to the developed chronological table.

The table of brick measurements presented here is still used in Warsaw to determine the approximate chronology of walls encountered in fieldwork in the city, though we are aware of the possibility of inaccuracies, but often have no other option, in the absence of other dating evidence.

An example of brick reuse in early seventeenth century Warsaw

Finally, I would also like to mention two interesting facts concerning the reuse of bricks in the buildings of Warsaw. The first one concerns the use of demolition brick, most likely from the city walls in the extension of the Church of Our Lady of Grace in the years 1758–1769 (by architect James Fontana). In one of the foundations, it was found that part of the wall face was constructed of Gothic bricks measuring 310–280 x 145–134 x 100–82mm.

The second example is of international significance.

---
[9] i.e. Rudkowski 1955, Miks 1957, Świechowski 1955, Wyrobisz 1961.

Figure 11. So-called 'Dutch' bricks of a yellow colour and of a small size (175 x 81 x 32mm, 180 x 77 x 36mm).

An interesting observation has been made during the investigations carried out in the past twenty years on several sites (the fortifications of the Royal Castle erected by the Vasa kings in 1620–1627; on the town's defensive walls in additions of the early seventeenth century 1621–1624; in several ecclesiastical and secular buildings, – for example, the Deacon's Palace by the Cathedral, rebuilt after 1610, and the buildings of the St Lazarus Hospital). During archaeological-architectonic work on these sites, fragments and sometimes whole examples of a different and rare type of brick, atypical for Warsaw have been found. These are examples of so-called 'Dutch' bricks of a yellowish fabric and small dimensions (175 x 81 x 32mm, 180 x 77 x 36mm) – (Figure 11). In the literature, it has been suggested that such bricks came to Poland from Western Europe in the ballast of ships and large boats transporting goods along the Vistula. When such vessels were loaded, the unneeded ballast was left on the river bank, and it could have been given to ecclesiastical and secular authorities. This hypothesis finds some support in the fact that in the first quarter of the seventeenth century, between 400 and 800 ships annually docked at Warsaw on their way down the river.[10]

The use of brick dimension analysis in other regions of Poland, research possibilities

The method presented here of the dating of historic walls by the measurements of the bricks has also had a much wider application in architectural and archaeological studies conducted in other Polish regions. In addition to their value in dating sites, brick measurements have also been used to identify and distinguish the production of construction workshops operating in central Poland (Wielkopolska).[11] The manufacture, dimensions and types of bricks used in the buildings of Western Pomerania have been studied by Antoni Kąsinowski.[12] Maria Brykowska has paid attention to the dimensions (and primarily the thickness) of bricks in her studies on the eastern extent of brick architecture in the culturally-differentiated area of the eastern margins of east-central Europe in the period from the thirteenth to sixteenth centuries.[13]

Conclusions

It seems that the method discussed here can be useful to apply to the study of buildings in regions where brick was the main construction material, above all in the differentiation of chronological phases of individual historical buildings and the identification and determination of the characteristics of local production centres. It however should be used only with extreme caution in comparison between structures over a wider area.

References

Baruch, M. and Wojciechowski, J. 1928. Kamienica książąt mazowieckich, Warszawa, Towarzystwo Miłośników Historii w Warszawie.

Bogucka, M. 1984. Podstawy gospodarcze rozwoju Warszawy. In M. Bogucka, M. I. Kwiatkowska, M. Kwiatkowski, W. Tomkiewicz and A. Zahorski (eds.), Dzieje Warszawy II, Warszawa w latach 1526–1795, Warszawa, Państwowe Wydawnictwo Naukowe (PWN), 43–79.

Brykowska, M. 2002. Studia nad wschodnim zasięgiem architektury ceglanej. In M. Arszyński and M. Mierzwiński (eds.), Cegła w architekturze środkowo-wschodniej Europy, Malbork, Muzeum Zamkowe w Malborku, 30–42.

Fischer, F. 1944. Norddeutsche Ziegelbau, München. Callway.

Gruszecki, A. 1965. Metoda graficzna badań pomiarowych cegły przy ustalaniu chronologii obiektów architektonicznych, Kwartalnik Architektury i Urbanistyki, X, no 1, 55–58.

Jabłczyńska-Jędrzejewska, H. 1958. Dawne zaprawy budowlane, Kwartalnik Architektury i Urbanistyki, X, no 1, 85–94.

Kąsinowski, A. 1970. Podstawowe zasady murarstwa gotyckiego na Pomorzu Zachodnim, Studnia z dziejów rzemiosła i przemysłu, 10, 47–131.

Kohte, J. 1895–1898. Verzeichnis der Kunsten mäler de Provinz Posen, Berlin. Verlag von Julius Springer.

Lloyd, N. 1925. A History of English Brickwork, London. H. Greville Montgomery.

Miks, E. 1957. W sprawie badań nad cegłą średniowieczną, Kwartalnik Historii Kultury Materialnej, V, no 1, 104–108.

Rudkowski, T. 1955. Badania nad rozmiarami cegły średniowiecznego Wrocławia, Sprawozdania Wrocławskiego Towarzystwa Naukowego, VII, 1952, supplement 5.

Schulze, K. W. 1927. Der Ziegelbau, Stuttgart. Dr. Frietz Wedekind &co

---

[10] Bogucka 1984, 46.
[11] Żemigała 2008.
[12] Kąsinowski 1970, 52–57.
[13] Brykowska 2002, 30–42.

Stiehl, O. 1898. Der Backsteinbau romanischer Zeit besonders in Oberitalien und Norddeutschland, Leipzig. Baumgärtner.

Stiehl, O. 1923. Backsteinbauten in Norddeutchland und Dänemark, Stuttgart. J. Hoffmann.

Stiehl, O. 1937. Der Backstein, Reallexicon zur deutchen Kunstgeschichte, Stuttgart. (J. B. Melzer).

Świechowski, Z. 1955. Architektura na Śląsku do połowy XIII wieku, Warszawa. Architektura i Budownictwo.

Tomaszewski, A. 1955. Badania cegły jako metoda pomocnicza przy datowaniu obiektów architektonicznych, Zeszyty Naukowe Politechniki Warszawskiej, Budownictwo, no 4, 31–52.

Wachtsmuth, F. 1925. Der Backsteinbau, Leipzig. Verlag der J. C. Hinrich' schen Buchhandlung.

Wachtsmuth, F. 1938. Der Backsteinbau der Neuzeit, Marburg. Elwertsche Universitäts- u. Verlagsbuchhandlung.

Widawski, J. 1970. Początki i rozwój murowanych obwarowań Warszawy przed epoką broni palnej, Kwartalnik Architektury i Urbanistyki, XV, 1970, no 3, 239–252.

Wyrobisz, A. 1961. Średniowieczne cegielnie w większych ośrodkach miejskich, Studia z dziejów rzemiosła i Przemysłu, 1, 55–82.

Żemigała, M. 2008. Cegła w budownictwie wielkopolskim w średniowieczu, Łódź, Wydawnictwo Instytutu Archeologii i Etnologii PAN.

FACTS AND THOUGHTS ON MEDIEVAL BRICK BUILDING IN THE NORTHERN PERIPHERY OF EUROPE

Gunilla Malm

Abstract: This article on medieval brick building in Sweden, in the northern periphery of Europe, focuses on the central area of the country, i.e. the area by the lakes Mälaren, Vänern and Vättern. It is based on many years of archaeological and building archaeological fieldwork carried out by the author as well as studies published by other scholars.

Brick as a building material was introduced in Sweden during the first half of the 13th century. Once brick was introduced, the material gained popularity. Builders using brick came from the social elite. Signs of work planning and work organization in brick masonry are difficult to find, but such signs do exist. Most probably, the Swedish medieval brick work forces were organized in guilds. Questions regarding costs and payment are discussed in the article. It concludes with facts and thoughts on the economy, aesthetic and ethical value of brick building.

Keywords: Aesthetic and ethical value, builder, bricklayer, Flemish double stretcher, guild, planning and organization, Vendish bond

Introduction

The purpose of the article is to give a short presentation on development and change in Swedish medieval brick building in the northern periphery of Europe, i.e. the area by the lakes Mälaren, Vänern and Vättern in Sweden. Those who initiated building with brick – the builders or initiators – and the bricklayers will be mentioned, as well as work planning, work organization and the guilds. The article will conclude with thoughts on the economic, aesthetic and ethical value of brick building.

Medieval Sweden extended over some 500 years, i.e. from the 11th century when the Viking Age ended to 1523 when Gustav Vasa became the Swedish king Gustav I. He was the king who, from a political, religious, socio-economic and geographic point of view, created the Sweden of today.

In the Swedish Middle Ages, the southern part of modern Sweden belonged to Denmark. In the west large areas belonged to Norway. The border towards Lappland in the north was vague. The coast area to the east of the Baltic Sea – the southwest of present day Finland – was part of Sweden´s territory. Gotland and Öland – the two Swedish islands in the Baltic Sea – sometimes belonged to Sweden, sometimes to Denmark, and sometimes they ruled themselves. In the centre of medieval Sweden are three lakes: Mälaren, Vänern and Vättern. This text focuses on the area around these lakes and refers to them and to the Baltic Sea when places with brick buildings are mentioned (Figure 1). Facts and thoughts presented in the text mainly concern medieval Sweden to the west of the Baltic Sea. Development and changes in brick building are a little bit different in the Swedish part to the east of the Baltic – i.e. part of modern day Finland.

Swedish brick building during the Middle Ages

The oldest brick buildings in Sweden are dated to the first half of the 13th century. Two of these buildings are churches to the north of Mälaren (Figure 1), i.e. the Dominican monastery church in Sigtuna and the Cistercian nunnery church in Sko.[1]

A squared brick tower in the Stegeborg castle to the south of Mälaren is another of these old brick buildings (Figure 1). Today it is a ruin. The Stegeborg castle, built of natural stone and brick, was added to the squared brick tower later, but still during the Middle Ages. Erik B. Lundberg has made an archaeological and building archaeological investigation of Stegeborg and studied the written material. He has put forward that the building of the tower was finished c. 1250, that the king was the builder and that it was meant for military use.[2] Because of technical building signs in the masonry I argue the tower is as old as Lundberg says, maybe somewhat older. Christian Lovén has questioned our interpretations.[3]

One additional building – a supporting or terrace wall – should be mentioned among the oldest brick buildings in medieval Sweden. On the ridge to the north of the Fyris River in Uppsala, located to the north of Mälaren (Figure 1), such a wall was once built towards the slope.[4] The building material was natural stone with buttresses of bricks. Nothing of that wall remains above ground today, but archaeological excavations have uncovered parts of it. It is preserved to a height of 2.4 m below ground level. We do not know how high it once rose above ground level. It

---

[1] Bonnier 1987, 26 f., 63 f., 163 ff.
[2] Lundberg 1964, 115 ff.
[3] Lovén 1999, 79 ff.
[4] Malm 1987a, 395; Anund 1992, 46.

Figure 1. Europe. Medieval Sweden with its vague borders. 1/ Mälaren. 2/ Vänern. 3/ Vättern. 4/ The Baltic.

was built on layers dated to between 1031–1232. Based on the layers, some finds and radiocarbon dates, Johan Anund has argued that this wall is as old as the above mentioned churches in Sigtuna and Sko.[5]

Building of the Archbishop´s Cathedral in Uppsala on top of the ridge started in the 1280s.[6] It meant the old wall had to be partly broken down otherwise it would block building the Cathedral. The decision to break the wall down was most probably taken at the same moment as the decision to build the cathedral on this site. Most probably, the king was the builder of the wall and also the one who decided to demolish it.[7]

Once brick began to be used as a building material in Sweden the material began to gain popularity. During the second half of the 13th century brick cathedrals were built, such as Strängnäs Cathedral to the south of Mälaren and Västerås Cathedral to the northwest of Mälaren. Some generations later Åbo/Turko Cathedral to the east of the Baltic Sea was built (Figure 1).[8]

At the end of the 13th century, some rural brick churches were also built. An architecturally strange rural brick church is in Tensta, located to the north of Mälaren (Figure 1) Most probably the builder was the archbishop Jakob Israelsson, which is why the church was likely founded between 1278 – 1281.[9]

The end of the 13th century was a period when royal palaces of a special kind were built. One is the royal brick palace in Vadstena to the east of Vättern (Figure 1). There are some other royal palaces in Sweden of this kind and from this time, but not of the same size. Behind the building of these palaces are influences from the continent where mighty kings and princes lived extremely luxurious lives. The royal brick palace in Vadstena is a mirror of this time and life, but is only a pale copy of the continental palaces.[10]

As stated above, Uppsala Cathedral was founded on the ridge to the north of the Fyris River in the 1280s. The builders of the Cathedral were the king and the archbishop. The building material was brick except for portal- and window-surroundings, pillars, decoration and so forth for which limestone was chosen.[11] Soon after the Cathedral´s foundation, the archbishop initiated building his brick manor to the west of the Cathedral. In medieval times, more buildings were added to the Cathedral area and soon a wall surrounded the complex. A Church Town was created – all of brick. Nowadays, Church Town walls and religious buildings have been demolished. But remaining buildings and walls, as well as old pictures and drawings, give a hint

---

[5] Anund 1992, 45.
[6] Dahlbäck 1977, 173 ff., Dahlbäck et al. 1984, 284 f.
[7] Malm in print.
[8] Bohrn et al. 1964 passim; Järpe 1979, passim; Medeltidsstaden 4 1977; Drake 2003a absract in English; Drake 2003b, abstract in English; Drake 2009, abstract in English; Drake 2011 passim; Lindroos et al. 2010 passim.
[9] Ramqvist 1996, 239 ff.
[10] Anderson 1972 passim; Jaubert 1993 passim.
[11] Boëthius and Romdahl 1935, 17ff.; Dahlbäck 1977, 173 ff.; Dahlbäck et al. 1984, 284 f.

of the former grandeur of the Church Town. The Cathedral is still the largest medieval brick stone monument in the north of Europe.[12]

According to a written note, Uppsala Cathedral was consecrated in 1435. The note and the date have been considered proof that building the Cathedral was finished, though some written notes tell that the building of the Cathedral was still going on decades after 1435.[13] If Uppsala Cathedral was founded in the 1280s and building works were finished in 1435 or some decades later, it means that building lasted for some 160 years. Generations after generations of Uppsala citizens never saw their Cathedral completed. Instead, year after year, they had to look at permanent scaffolds standing. But maybe this view was not remarkable. During this time grand cathedrals were built in many towns in Europe.

Of course, Uppsala Cathedral has been the subject of many investigations and publications. One of the latest is Göran Dahlbäck's article on signs of brick production in Uppsala Cathedral during the end of the Middle Ages. In 1210, Sveriges Kyrkor (Swedish Churches) published a grand work on Uppsala Cathedral.[14]

Most probably, the building of Uppsala Cathedral and its Church Town gave rise to a wider use of brick in building in Sweden. New brick monuments were built and those already built of natural stone were added to with parts in brick such as vaults, decorations, portal- and window-surroundings. Churches were added to with, for instance, new brick vestries, towers and chapels.[15] Some castles had brick masonry added on their surface of natural stone masonry, as if the brick masonry were some kind of "wall tapestry" (Figure 2).

A good example of addition and change with brick as a building material is the Holy Trinity Church in Uppsala some 100 meters to the south of the Cathedral and the Church Town. The building material is natural stone and brick. The foundation of this church dates to the end of the 13th – beginning of the 14th century. Later on, but still in the Middle Ages, a tower, a church porch and chapels – all of brick – were added to the aisle of natural stone.

The currently standing Holy Trinity Church in Uppsala was built on the very spot of an older Holy Trinity. That Holy Trinity was most probably a wooden church. We do not know the age of that church, but written material and the results of archaeological and building archaeological investigations indicate it was standing on this very site in

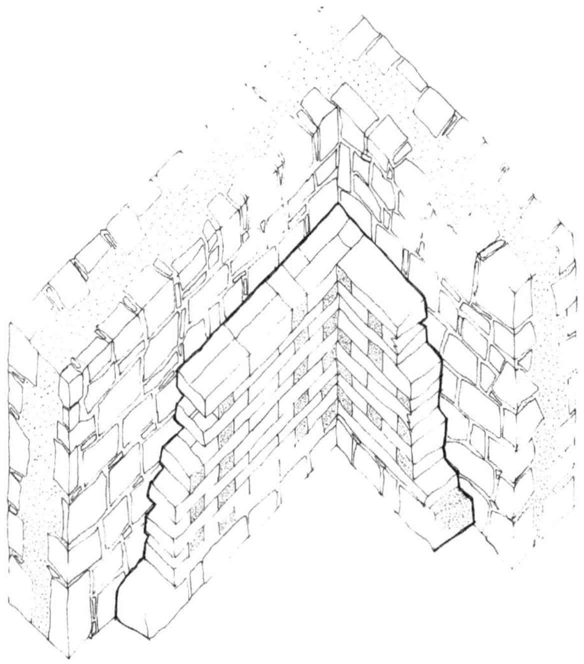

Figure 2. Sketch of brick masonry added onto natural stone masonry as if the brick masonry is some kind of a "wall tapestry".

the 12th century.[16] Recently though, new interpretations have been presented.[17]

Military castles – military monuments – with parts built of either brick or natural stone were initiated by the king or the archbishop in the Middle Ages. For instance, kings were the builders of such castles in Västerås and Tavastehus. The former is located northwest of Mälaren, the latter east of the Baltic, i.e. in modern day Finland (Figure 1). The archbishop was the builder of such a castle in Uppsala.[18] It was located to the west of the older archbishop's manor, named above.

The urban world must be mentioned when discussing brick building in the Middle Ages. It should be noted that the buildings in most towns by then were wooden. Except for natural stone and brick churches, just a few had brick or natural stone buildings, such as for instance the towns of Stockholm by Mälaren, Uppsala to the north of Mälaren, Vadstena to the east of Vättern, Visby on the island of Gotland in the Baltic Sea, and Åbo/Turku to the east of the Baltic Sea (Figure 1).[19] During this period many towns built town churches of brick such as Sankt Lars in Söderköping,

---

[12] My colleague and mentor in building archaeological and antiquarian questions was Dr Åke Nisbeth, Central Board of National Antiquities, Sweden. I remember him often saying that Uppsala Cathedral and the Church Town once was the only Swedish monument comparable to continental monuments according to largeness and dignity.
[13] Boëthius and Romdahl 1935, 165, 192; Dahlbäck et al. 1984, 285.
[14] Dahlbäck 2008 passim; Sveriges Kyrkor 2010, volume 229–232 passim.
[15] Boëthius 1918 passim; Boëthius 1921 passim; Bonnier 1987 passim.

[16] Malm 1987b passim; Malm in print.
[17] Sveriges Kyrkor volume 228 1210, 80 f., 87, 187 ff., 347 ff.
[18] On the castle in Västerås: Lovén 1999, 175 f; Nordberg 1975, 30 ff; On the castle in Tavastehus: Drake 1968 esp. 37 ff; Drake 2001a, 212.217; Drake 2001b, 123–128; Drake 2003 c, 11–14; Lovén 1999, 94 ff. On the castle in Uppsala: Hahr 1929, 151; Lovén 1999, 258 ff.
[19] On medieval stone houses in Turku see Uotila 2003, English abstract; Uotila 2007, English summary. See also Ratilainen 2010 passim and Ratilainen 2012 passim.

Sankt Per in Vadstena and Vårfrukyrkan in Skänninge,[20] all to the south of Mälaren (Figure 1).

A new group of buildings that emerged during the second half of the Swedish Middle Ages are cellars built of natural stone or brick, i.e. storage areas, shelters or safe rooms used for keeping the merchants´ costly or expensive goods. These cellars are most often seen in towns, but they are also a rural phenomenon.[21]

Finally, it should be underlined that brick never replaced natural stone as a building material. Examples of large natural stone buildings are the cathedrals in Skara founded in the 12th century and Linköping founded in the 13th century. Both Cathedrals have major natural stone parts added during later medieval building periods.[22] Skara is located between the lakes Vänern and Vättern, Linköping to the south of Mälaren (Figure 1). Further on we have the natural stone abbey church in Vadstena to the east of Vättern (Figure 1). This church was founded in the late 14th century.

The builders – those who initiated brick as building material

Those who initiated building in brick in medieval Sweden always came from the social elite. The king, members of the royal family or members of the church introduced brick as a building material and were the initiators of building cathedrals, churches, palaces and castles.

Gradually the society and socio-political life changed and soon we also see builders from a different social class, i.e. wealthy merchants or those connected to trade, as well as other kinds of powerful citizens of the towns. These builders may have been related to the king, the royal family or the religious world, but not necessarily. They belonged to the elite of the society by virtue of their wealth.

It should be noted that women were also builders.[23] One representative of this group is Birgitta Ulfsdotter, canonized and remembered in the modern era as Holy Birgitta. She was related to the royal family. Her intention to establish an Abbey was realized when the Swedish king Magnus Eriksson and his queen Blanka donated their royal brick palace in Vadstena to her in 1346 with the purpose of rebuilding it into her abbey. Written material tells us that Birgitta Ulfsdotter was a powerful initiator of a grand building und she had a great technical talent for building. She gave clear directions on how to change the royal brick palace to fit her abbey's intensions and how to build the church to give it the shape she wanted. She cited precise wishes for the building material of the abbey church – such as natural stone must be used as building material except for the vaults. The vaults must be the only part of the church built of brick.[24]

Birgitta left Sweden before the building of the abbey in Vadstena started. She died in Rome and her body was taken to her homeland shortly thereafter. It is said a route via the island of Öland in the Baltic Sea was chosen, as recently described by Marie-Louise Sallnäs.[25] During the end of the Middle Ages, Birgittine daughter convents were established in Denmark, Germany, Estland, and in Nådendal in modern Finland to the east of the Baltic Sea (Figure 1). In these monuments, brick was used. In other words, Birgitta´s directives on building material were not followed.[26] The Birgittiner Convent in Nådendal has recently been reinvestigated and the investigation has been published.[27]

The bricklayers

When building the oldest brick monuments in Denmark in the mid and second half of the 12th century, the brick-working force had to be invited from abroad.[28] But who built the first brick monuments in Sweden some generations later? Whoever it was, they knew how to produce bricks and build brick masonry with regular rhythms according to medieval praxis.[29] Maybe these brick-working forces came from abroad, maybe from Denmark after having finished their job there, or maybe they were Swedish. Most probably Sweden gradually developed brick-working forces of its own. But it should be remembered that members of brick-working forces belonged to an ambulating profession. A brick layer could finish one job in Uppsala and start another in Västerås. He could move from a job in Sweden to Denmark or Germany.

Medieval bricklayers are invisible like many workers from any time. Their names are seldom found in written material. But it does happen. A letter dated the 11th of April, 1366 from Uppsala Cathedral mentions "murator" Magnus Nilsson and his servants Herman, Ulf and Magnus. We hear of still another "murator" in Uppsala Cathedral – a Thyrgillos murator. The 10th of February, 1430 Diarium Vadstenense tells us as follows: "sepultus fuit hic quidam servitor monasterii nomine Thyrgillos murator, qui aliquanto tempore demoratus apud ecclesiam upsalensem I aertificio suo, modo relatus est mortuus".[30]

On signs of bricklayers in brick masonry, see below.

Bricklayers behind the bonds

During the Swedish Middle Ages, there are two different brickwork bonds, namely Flemish double stretcher and

---

[20] Broberg and Hasselmo 1978 passim; Hasselmo 1982 passim, Malm 1983 passim; Hasselmo 1984 passim.
[21] Andersson and Redin 1980, 33, 35; Ersgård 1986, 94; Redin 1976, 37.
[22] Skara domkyrka, www.vastsverige.com, Cnattingius et al. 1987, 242 ff.
[23] Wienberg 1997, 53.
[24] Anderson 1991, 10, 54.
[25] Sallnäs 2001 passim.
[26] Berthelsson 1947, 507–521 passim.
[27] Uotila 2011a passim; Uotila 2011b, 29–37; Uotila 2011c, 67–74; Uotila and Väisänen 2011, 227–238; Uotila et al. 2011, 303–307.
[28] Tuulse 1968, 53.
[29] On brick sizes, bond technique and proportions see: Sundnér 1982, 53.62, 68–89 and the summary pages 107–126.
[30] Boëthius and Romdahl 1935, 160.

Vendish bond, the first one with two different rhythms, the last one with one rhythm only (Figure 3 A, B, C). Upwards of that, we have an abundance of irregular bonds.

Flemish double stretcher and Vendish bond exist throughout the Middle Ages. In Sweden, to the west of the Baltic Sea, there is no doubt though that the Flemish double stretcher is more often used during the beginning of the era and the Vendish bond more often during the later part. For instance, we have Flemish double stretcher in the old churches in Sigtuna and Sko, as well as in the brick tower of Stegeborg. To continue, Flemish double stretcher has been used in the oldest building periods at the royal palace of Vadstena and the cathedrals of, for instance, Uppsala and Västerås. In the later building periods of the two cathedrals, Vendish bond or irregular rhythms have been used. The reason for the change of bonds is unknown, see thoughts and discussion below.

Among Swedish brick scholars, it has been argued that the use of the different rhythms is because of different technical reasons, for instance that the Vendish bond gives a more solid masonry due to the increased use of binders. Perhaps this is correct, but I will argue there might also be other explanations.

Söderköping is a little town to the south of Mälaren and to the east of Vättern (Figure 1). Here we have the so-called "Bishop Brask´s printing office", a brick building from the 15th century with a later natural stone addition. A restoration and a building archaeological investigation took place in 1989. When plaster was removed from the south façade, both rhythms of Flemish double stretcher and an irregular bond could be clearly seen (Figure 4); on the drawing of the façade you have Flemish double stretcher to the right (Figure 3, A), a somewhat irregular Flemish double stretcher in the middle (Figure 3, B) and an irregular bond to the left. When the rhythms are found side by side like that, I argue that they mirror other than technical matters. I argue the rhythms represent three different brick layers working side by side – teamwork so skillfully made that you cannot see any joints between their rhythms when the work is finished. [31]

My interpretation of the different rhythms of the south façade of course raises questions such as: does a bricklayer always use one and the same bond? Without having any proof, I argue "yes" and that because of a special reason. To me, raising brick masonry with regular rhythm is a most complicated handicraft. It must take years to be skillful at one bond and once you are skillful you use that bond the rest of your working life. I think masters are teaching their journeymen one of the rhythms – probably the rhythm the master is good at himself. The journeymen are using that rhythm the rest of their lives – that is, if nothing else is ordered because of technical, decorative or other reasons.

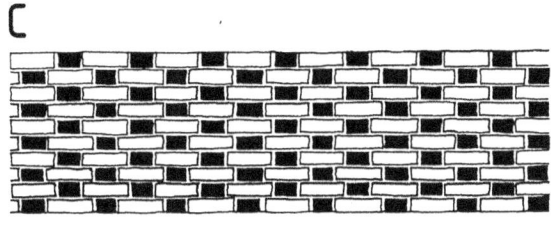

Figure 3. Medieval bonds. A – B = Flemish double stretcher, i.e. two stretchers followed by a header in two different rhythms. C = Vendish bond, i.e. one stretcher followed by a header. In: Malm 2000, 206.

Signs of planning and organization

In the masonry of Uppsala Cathedral you can see that the building work has breaks – different building phases can be separated. For example, above the vault of the crossing, the masonry has been built to a certain height with a certain bond and then building has come to a break. When the work continues another bond has been used,[32] i.e. according to my interpretation different bricklayers have started, continued and ended the bricklaying work. I argue that this not only mirrors different bricklayers but also the work leader planning and organizing his working staff – a planning and organizing influenced by technical building reasons, economy, available working time and working material.

The age of the squared brick tower in Stegeborg – facts and thoughts on its bond, brick size and lack of scaffolding holes

Above I argued that the squared brick tower of Stegeborg is one of the oldest brick buildings in Sweden. My interpretation is based on the bond, Flemish double stretcher (Figure 3, B), the large size of the bricks and on the lack of scaffolding holes. As stated above, Flemish double stretcher is in use throughout the Middle Ages but is preferred at the beginning of this period.

---

[31] Malm 1992 passim; Malm 2000 passim.

[32] Malm 1998 ATA, RaÄ.

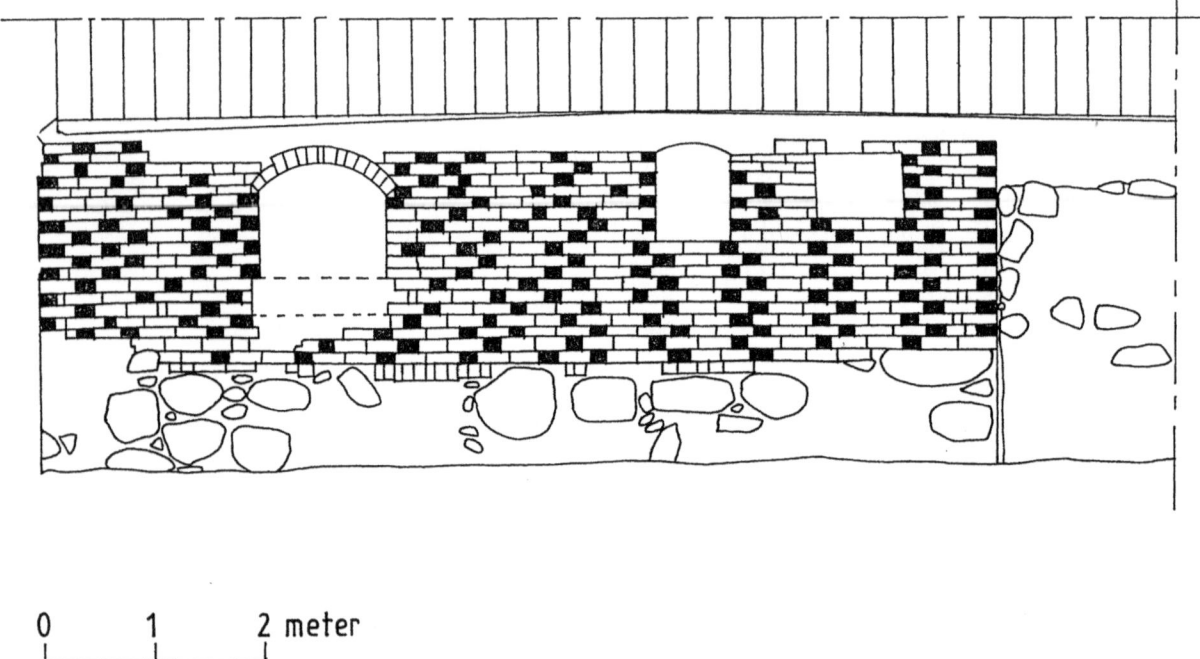

Figure 4. The so-called "Biskop Brask's printing office" in Söderköping. South facade. The header of the bricks is marked with black. Scaffold holes not visible. In: Malm 2000, 207.

According to (unpublished) investigations the size of bricks is becoming more standardized during the second half of the Swedish Middle Ages.

During the first half of the period variations of size is more often found. For instance, there are bricks with a large size as in the squared brick tower in Stegeborg – a size that is rare during the second half of the period. In one of the oldest building phases of Uppsala Cathedral the bricks have a small size – a size not found in the later building phases – a size that is quite rare during the second half of the Swedish Middle Ages.

In the building phases of Uppsala Cathedral from the 14th and 15th centuries, the size of the bricks is almost the same though they most probably are coming from different brick producers. This size seems to be similar to the brick size in other brick buildings from this period of the Middle Ages – apparently a standardized size.

Furthermore, the masonry of the brick tower of Stegeborg does not have scaffolding holes. A rare kind of scaffold must have been used. [33] Also, scaffolds seem to have been more standardized during the second half of the Swedish Middle Ages. For instance, the distance between the put log holes both vertically and horizontally are almost the same.

The bond Flemish double stretcher, the large brick size and the lack of scaffolding holes in the squared brick tower of Stegeborg are something found in brick buildings throughout the Swedish Middle Ages, most often in the oldest period of those Ages, though and by then, most often only one or two of the variants is found in one single building. Finding all three variants like in the brick tower of Stegeborg gives a hint of the older-fashioned brick building. Hence, I will argue that the building is one of the oldest in Sweden.

Brick building and guild

At the end of the 13th – mid-14th century, the first signs of guilds in Sweden are found – the mason's marks of Linköping Cathedral.[34] The marks reflect some kind of organization dealing with the social, economic and juridical interest of the working force. But mason's marks are linked to natural stone buildings. [35] I know of only one brick building having mason's mark, namely the exterior wall of All Saint's Chapel in Åbo Cathedral.[36] To Juhani Rinne these marks rather are brick maker's marks and some kind of control marks. [37]

Sometimes prints of, for example, animal paws, children's feet or fingers can be seen on bricks. Bricks sometimes also have strokes, crosses, dots, stars and other figures of that kind. Marks of this kind can be seen in bricks from the Abbey building period in the former royal palace in Vadstena. These marks are hard to interpret. Are they scrawling, control marks or some kind of brick maker's marks?[38]

---

[33] On medieval scaffolds, See: Ratilainen http://utu.academia.edu/TanjaRatilainen
[34] Cnattingius et al. 1987, 87.
[35] Cinthio 1957, 87; Malm 2001 passim.
[36] Tanja Ratilainen kindly has drawn my attention to these marks.
[37] Rinne 1941, 309.
[38] Anderson 1972, 174 ff.

Despite the lack of mason´s marks on bricks, most probably the medieval brick working force was organized into a guild, i.e. they had the benefit of the advantages organizations like guilds could offer according to socio-economic and juridical circumstances.

Facts and thoughts on the economic, aesthetic and ethical value of brick work during the Swedish Middle Ages

Let me finish my text with facts and thoughts on the economic, aesthetic and ethical value of medieval brickwork during the Swedish Middle Ages.

Economic value

As mentioned above, in the course of the Swedish Middle Ages, Vendish bond gradually grew more popular than the Flemish double stretcher in Sweden west of the Baltic Sea, though both bonds were in use throughout the Middle Ages. We do not know the reason for the change, but the change seems to coincide with the first mason´s marks on natural stone architecture, i.e. the first signs of a guild. Such circumstances bring up the question of if the change depended on matters or negotiations between builder and work force concerning available time, quality of the work and payments.

Vendish bond demands a larger amount of bricks than Flemish double stretcher. If brick had an economic value, using Vendish bond would stipulate a higher price than using Flemish double stretcher. Why, then, was the Vendish bond preferred? Was the Flemish double stretcher more complicated – did it take a longer time using that bond – was working time more economically valuable than the bricks?

Was brick producing or brick building expensive or cheap? Did brick building at the end of the Swedish medieval period have a fixed price or was the price subject to negotiations? Let me ask the questions in another way.

At the end of the Swedish medieval Ages, brick building was going on at Uppsala Cathedral, at the brick castle of Västerås and at brick cellars in some Swedish towns – i.e. different working places, different bricklayers and different working forces. Did a bin of brick for the Cathedral, a bin of brick for the castle and a bin of brick for a cellar command the same or different prices?

Aesthetic value

As mentioned above, some medieval castles have brick masonry added onto the surface of natural stone masonry as if the brick masonry was some kind of a "wall tapestry" (Figure 2). This phenomenon may be seen, for instance, in the castles of Stegeborg and Västerås as well as in the castle of Kronoberg quite far to the south of Mälaren (Figure 1). As I understand, the surface of the natural stone masonry did not have any need of support or restoration. In other words, I cannot find any other reason for this way of brick building than that the brick "tapestry" is added to the natural stone surface for aesthetic reasons – the owner of the stone building preferred looking at brick masonry than masonry of natural stone.[39]

Ethical value

The royal brick palace of Vadstena was built in the last half of the 13th century with the king as the builder. Birgitta Ulfsdotter became the creator and builder of the abbey some 75 years later when the royal palace was given her as a gift. Birgitta was award of the former life of luxury and extravagance at the royal brick palace – a life far from the humble life she wanted for herself or the abbey. She was looked upon as an ascetic person. Probably to her, brick symbolized an exclusive and luxurious life, a life she could not accept. The reason Birgitta rejected brick as a building material in the abbey church, except for the vaults, may have been that the material symbolized extravagance to her and her wish was to give the church a look of simplicity – at least according to written material.[40]

Summary

Brick as building material was introduced in Sweden during the first half of the 13th century. The new building material soon became popular. Cathedrals, palaces, castles, urban and rural churches were built – all buildings connected to the social elite of the society. From the second half of the Swedish Middle Ages brick and stone cellars were constructed by the elite. Brick vaults, decorations, portal- and window-surroundings were added in many stone buildings during the same period. Churches received new brick vestries, towers and chapels.

The king, the royal family or members of the church were the first initiators of brick building projects. Gradually the society changed, and so wealthy merchants as well as other powerful members of town communities were acting as builders. They belonged to the elite of the society by virtue of their wealth. Powerful woman were also builders. Contrary to these mighty builders, the bricklayers are quite invisible but sometimes brick bonds mirror their participation or handicraft.

In brick masonry, signs of planning and organizing are difficult to find, but they do exist. Changes and interruptions in a brick bond mirror different bricklayers´ and the work leaders´ planning and organizing of the work. Mason´s marks mirror the existence of a guild. Despite the lack of mason´s marks in brick masonry most probably the medieval brick work forces were organized in guilds.

It is not known why the Vendish bond gradually became more popular than the Flemish double stretcher, but it seems to coincide with the first mason´s marks on natural stone

---

[39] On aesthetic or economic value see also Lovén 1999, 175 f.
[40] Anderson 1991, 10 ff.

architecture, i.e. the first signs of guilds. Such circumstances bring us to the question of if the change depended on the negotiations between the builder and work force concerning available working time, quality and economy.

In some cases, aesthetic reasons or symbolic value affected the decision to use brick. We have examples in which stone walls without obvious need to repair were covered with brickwork. It is also possible that Birgitta Ulfsdotter refused to use brick because it symbolized extravagance to her.

Acknowledgements

I would like to thank Tanja Ratilainen for good companionship during my writing process of this article. Many thanks to Jean Price who proofread the article.

References

Anderson, I. 1972. Vadstena gård och kloster. Stockholm, Kungliga Vitterhets Historie och Antikvitets Akademien.

Anderson, I. 1991. Vadstena klosterkyrka I. Kyrkobyggnaden. Med bidrag av Sune Ljungstedt and Gunilla Malm. Sveriges Kyrkor volume 213. Stockholm, Riksantikvarieämbetet, Kungliga Vitterhets Historie och Antikvitets Akademien.

Andersson, H. and Redin, L. 1980. Medeltidsstaden 19. Stadsarkeologi i Mellan-Sverige. Läge, problem, möjligheter. Göteborg, Riksantikvarieämbetet och Statens historiska museer.

Anund, J. 1992. Domkyrkoplan i Uppsala. Arkeologisk schaktövervakning 1992. Riksantikvarieämbetet och Statens historiska museer. Rapport. UV 1992:7. Uppsala.

Berthelsson, B. 1947. Studier i Birgittinerordens byggnadsskick I. Anläggningsplanen och dess tilllämpning. Enlish summary, 507–521. Kungliga Vitterhets, Historie, och Antikvitets Akademiens Handlingar, del 63. Lund.

Boëthius, G. 1918. Stilströmningar i den uppländska tegelarkitekturen. Studier i Upplands kyrkliga konst, 175–190, Stockholm.

Boëthius, G. 1921. De tegelornerade gråstenskyrkorna i norra Svealand. Ett bidrag till stilströmningarna under den yngre medeltiden. Studier från Stockholms högskolas konsthistoriska institut 3 Stockholm, Stockholms högskola.

Boëthius, G and Romdahl, L. A. 1935. Uppsala dokyrka 1258 – 1435. Uppsala, Almqvist & Wiksell.

Bonnier, A. C. 1987. Kyrkorna berättar. Upplands kyrkor 1250 – 1435. Upplands fornminnesförenings tidskrift 51. Uppsala, Utgiven av Uppland Fornminnesförening och Hembygdsförbund.

Bohrn, E., Curman, S. and Tuulse, A. 1964. Strängnäs domkyrka medeltida byggnadshistoria. Text. Band I. Häfte I. Konsthistoriskt inventarium. Kungliga Vitterhets Historie och Antikvitets Akademien. Sveriges Kyrkor volume 100.

Broberg, B., and Hasselmo, M. 1978. Medeltidsstaden 5. Söderköping. Göteborg, Riksantikvarieämbetet och Statens historiska museer.

Cinthio, E. 1957. Lunds domkyrka under romansk tid. Acta Archaeologica Lundensia I. Lund, Lunds universitet.

Cnattingius, B., Edenheim, R. and Ullén, M. 1987. Linköpings domkyrka. I. Kyrkobyggnaden. Sveriges Kyrkor volume 200. Stockholm, Almqvist & Wiksell International.

Dahlbäck, G. 1977. Uppsala domkyrkas godsinnehav med särskild hänsyn till perioden 1344 – 1527. Studier till det Medeltida Sverige 2. Stockholm, Kungliga Vitterhets Historie och Antikvitets Akademien, Almqvist & Wiksell Internationell.

Dahlbäck, G., Ferm, O. and Rahmqvist, S. 1984. Det Medeltida Sverige 1:2. Tiundaland: Ulleråker, Vaksala, Uppsala stad. Stockholm, Riksantikvarieämbetet.

Dahlbäck, G. 2008. Uppsala domkyrkas tegeltillverkning vid medeltidens slut. In G. Dahlbäck, C. Ekholst, B. Eriksson, G. Bjarne Larsson, S. Nyström and M. Åkestam (eds), Medeltidens mångfald. Studier i samhällsliv, kultur och kommunikation tillägnade Olle Ferm på 60-årsdagen den 8 mars 2007, 56–67. Stockholm, Sällskapet Runica et Mediaevalia.

Drake, K. 1968. Die Burg Hämeelinna. Eine baugeschichtliche Untersuchung. Helsingin yliopisto.

Drake, K. 2001a. Die Bauherren der Burg Häeenlinna im Mittelalter. In Castella Maris Baltici 3–4, AMAF V; Alttoa K.; Drake, K.; Pospiezny, K and Uotila. K. (eds), 212–217. Turku.

Drake, K. 2001b. Die Deutschordensburgen als Vorbilder in Schweden. In Burgen kirchlicher Bauherren. Volume 6, 123–128. München, Berlin.

Drake, K. 2003a. Åbo domkyrka och byggnadsarkeologin. In Kaupunkka pintaa syvemmältä, Archeologia Medii Aevi Finlandiae IX, Liisa Seppänen (ed), 135–152. Suomen keskiajan arkeologian seura, Turku.

Drake, K. 2003b. Åbo domkyrkas äldre byggnadshistoria. In Songs of Ossian. Festschrift in honour of professor Bo Ossian Lindberg, 83–92. Helsinki 2006.

Drake, K. 2003c. Häme castle as a subject of research. In At home within Stone Walls. AMAF VIII. Vilkuna A-M. and Mikkola. T. (eds), 11–14. Suomen Keskiajan arkeologian seura. Saarajärvi.

Drake, K. 2009. Åbo gråstensdomkyrka. In Earth, Stone and Spirit. Markus Hiekkanen festschrift. Hanna-Maria Pellinen (ed), 182–191. Soumen keskiajan arkeologian seura. Saarijärvi.

Drake, K. 2011. The First Conception of Masonry Cathedral of Turku. In Times, Things and Spirit. 36 essays for Jussi-Pekka Taavitsainen, 372–379.

Ersgård, L. 1986. Expansion och förändring i Uppsala medeltida bebyggelse. In Från Östra Aros till Uppsala. En samling uppsatser kring det medeltida Uppsala. Uppsala stads historia volume 7, 78–100. Uppsala.

Hahr, A. 1929. Uppsala forna ärkebiskopsgård. Sankt Eriks gård och dess historia. Uppsala.

Hasselmo, M. 1982. Medeltidsstaden 36. Vadstena. Göteborg, Riksantikvarieämbetet och Statens historiska museer.

Hasselmo, M. 1984. Medeltidsstaden 40. Skänninge. Göteborg, Riksantikvarieämbetet och Statens historiska museer.

Jaubert, A. N. 1993. Prince's Residences in Denmark – from ca 1000 to ca 1350. In K. Drake (ed.), Castella Maris Baltici I, Archaeologia Medii Aevi Finlandiae 1, 89–100. Stockholm, Almqvist & Wiksell.

Järpe, A. 1979. Medeltidsstaden 10. Strängnäs. Göteborg, Riksantikvarieämbetet och Statens historiska museer.

Lindroos, A.; Ringbom, Å.; Kaisti, R.; Heinemeier, J. and Brock, F. 2010. The Oldest Parts of Turku Cathedral. C.14 Chronology of Fire Damaged Mortars. In Archaeology and History of Churches in Baltic Region Symposium, June 8–12, 2010, Visby, Sweden.

Lovén, C. 1999. Borgar och befästningar i det medeltida Sverige. Stockholm. Kungliga Vitterhets Historie och Antikvitets Akademien.

Lundberg, E. B. 1964. Stegeborgs slott. Kungliga Vitterhets historie- och antikvitetsakademiens handlingar. Antikvariska serien 12. Stockholm.

Malm, G. 1983. Sockenkyrka – stadskyrka. META Medeltidsarkeologisk tidskrift 2.

Malm, G. 1987a. Recent excavations at Uppsala Cathedral, Sweden. World Archaeology. Vol. 18 No 3, 382–397.

Malm, G. 1987b. Helga Trefaldighets kyrka i Uppsala. Årsbok för medlemmarna i Upplands Fornminnesförening och Hembygdsförbund 1985/86, 7–24.

Malm, G. 1992. Medeltidens murade byggnader som arkeologiska artefakter. In Medeltida husbyggande. Symposium i Lund 1989. Lund Studies in Medieval Archaeology 9, 221–248. Stockholm.

Malm, G. 1998. Rapport över arkeologiska och byggnadsarkeologiska undersökningar i Uppsala domkyrka. Unpublished report. Ritningar (Drawings). Antikvarisk topografiska arkivet (ATA). Stockholm, Riksantikvarieämbetet (RaÄ).

Malm, G. 2000. Murarna berättar. In Eriksdotter, G., Larsson, S., and Löndahl, V. (eds.) Att tolka stratigrafi. Det tredje nordiska stratigrafimötet. Åland. 1999. Meddelande från Ålands högskola 11, 204 – 210. Mariehamn.

Malm, G. 2001. Vadstena Abbey church and its Mason's marks. In Archaeology and Buildings. Papers from a Session held at the European Association of Archaeologists Fifth Annual Meeting in Bournemouth 1999. British Archaeological Reports International, 57–85. Oxford.

Malm, G. Ground planning for building Uppsala Cathedral. In print.

Medeltidsstaden 4. Västerås. 1977. Göteborg, Riksantikvarieämbetet och Statens historiska museer.

Nordberg T. O:son. 1975. Västerås slott. En byggnadshistorisk skildring. Västerås kulturnämnds skriftserie 2. Västerås, Västerås kommun.

Rahmqvist, S. 1996. Sätesgård och gods. De medeltida frälsegodsens framväxt mot bakgrund av Upplands bebyggelsehistoria. Upplands fornminnesförenings tidskrift 53. Uppsala, Upplands fornminnesförening.

Ratilainen, T. Quest for Medieval Masons and the Building Techniques They Applied. http://utu.academia.edu/TanjaRatilainen

Ratilainen. T. 2010. Archaeology and History of Churches in Baltic Region Symposium, June 8–12, 2010, Visby, Sweden.

Ratilainen, T. 2012. To Study Medieval Brick Building Technique by Applying Digital Surveying Methods Cases: Holy Cross Church of Hattula and Häme Castle. In Castella Maris Baltici X. Uotila, K.; Mikkola, T. and Vilkuna, A-M. (eds), 197–208. Saarijärvi.

Redin, L. 1976. Medeltidsstaden 3. Uppsala. Göteborg, Riksantikvarieämbetet och Statens historiska museer.

Rinne, J. 1941. Turun tuomiokirkon historia I, Tuomiokirkon rakennushistoria. Turko, Turon tuomiokirkon isännistö.

Sallnäs, M. L. 2002. Den heliga Birgittas hemkomst. Tradition och verklighet i öländskt tretonhundratal. Kalmar län 2002, årsbok för kulturhistoria och hembygdsvård 86, 22–34.

Skara domkyrka, www.vastsverige.com

Sundnér, B. 1982. Maglarp, en tegelkyrka som historiskt källmaterial. Lund.

Sveriges Kyrkor 2010, volume 227–231. Uppsala domkyrka I–VI. Uppsala. Kungliga Vitterhets Historie och Antikvitets Akademien.

Tuulse, A. 1968. Romansk konst i Norden. Stockholm.

Uotila, K. 2003. Kivitaloja keskiajan Turussa. In Kaupunkia pintaa syvemmältä. Arkeologisia näkökulmia Turun historiaan. AMAF IX. Liisa Seppänen (ed.), 121–134. Suomen keskiajan arkeologian seura, Turku.

Uotila, K. 2007. ABOA VETUS –museon kivirakennusten tutkimukset v. 2002–2006. SKAS (Sällskapet för medeltidsarkeologi i Finland. 2007:2), 18–27.

Uotila, K. (ed.) 2011a. Naantalin luostarin rannassa. Stranden vid Nådendals kloster. Kaarina, Muuritutkimus.

Uotila, K. 2011b. Nådendals äldsta historia och dess forskare. In K. Uotila (ed.), Naantalin luostarin rannassa. Stranden vid Nådendals kloster, 29–37. Kaarina, Muuritutkimus.

Uotila, K. 2011c. Klosterbyggandets historia. In K. Uotila (ed.). Naantalin luostarin ramassa. Stranden vid Nådendals kloster, 67–74. Kaarina, Muuritutkimus.

Uotila, K., and Väisänen, T. 2011. Utgrävningsområdena och fynden. In K. Uotala (ed.), Naantalin luostarin rannassa. Stranden vid Nådendals kloster, 227–238. Kaarina, Muuritutkimus.

Uotila, K., Helama, S., Hotopainen, J., Hilasvuori, E., and Oinonen, M., 2011. Byggandet av Nådendals kloster under 1400-talet. In K. Uotila (ed). Naantalin luostarin ranassa. Stranden vid Nådendals kloster, 303–307. Kaarina, Muuritutkimus.

Wienberg, Jes. 1997. Stormänd, stormandkirker og stormandsgårde – kritiske reflektioner. META Medeltidsarkeologisk tidskrift 4.

A Recently Excavated Medieval Brick Kiln at Saltsjö Boo, Sweden and the Hxrf Analyses of its Products

Jan Peder Lamm and Anders Lindahl

Abstract: The brick kiln at Saltsjö Boo is one of very few such structures known from medieval Sweden. Approx ¾ of the kiln has been excavated in successive short excavations from 2006 to 2011. Carbon 14 and TL (Thermoluminescence) analyses indicate that the kiln has been in operation from c. 1250 to c.1450. This paper presents the results of the excavations and discusses the production of bricks and industrial logistics of the site.

To study brick manufacture and changes that occurred in composition of the raw material during over a period of 200 years we have used a handheld XRF, which offers great possibilities for chemical analyses on ceramics. A clear advantage of the method is that a large amount of samples may be analysed at low cost and during a short period of time. The data acquired at the analysis have been processed by means of various statistical methods from interpretations of scatter plots of two elements at time to multivariate tests. Of special interest has been to determine what may be considered as a normal variation in a sample from this type of context and explain anomalies in the material.

Keywords: Archbishop, Birger Jarl, bricks, brick-and lime kiln, brick production, clay, Conradus de Arxö, Franciscan order, medieval kiln, Tideman Fries

Introduction

In a series of short summer campaigns 2006–2011 a medieval brick kiln, also used for making lime, was found and partly excavated at Saltsjö Boo in Nacka municipality c. 15 km east of Stockholm. A full excavation report was printed in 2013.[1] Very few such structures are known in Sweden, fewer have been archaeologically investigated and even fewer published. Thus the opportunity is now taken to introduce this important site to an international audience.

In Denmark the situation is quite different. There brick kilns are considered to be the most common type of medieval building remains.[2] In Norway and Finland, however, kilns are even rarer than in Sweden: three have been found in the Oslo fjord area[3] and a fourth at Kungahälla in former Norwegian Bohuslän (now in Sweden).[4] The only Finnish kiln of medieval date is located in Hattula, Herniäinen.[5]

The foundations of our kiln are solidly constructed and astonishingly well preserved. According to the radiocarbon dating results it was built in the mid-13th century cal. AD According to radiocarbon and thermoluminescence analyses it may still have been in use in the early 16th century.[6] The Boo kiln has close parallels in Denmark which aid in the interpretation of construction and capacity.

One of the main objectives in this presentation has been to study the variation in the chemical composition of the clay which was used to manufacture bricks in the Boo kiln.

Early brick buildings in the Stockholm area

It is commonly held that in the early 12th century the art of making brick saw a renaissance in Northern Italy. The rediscovered technology then spread quickly. In about 1160 it reached Denmark[7] and soon the new building technique became fashionable also farther north.

Around Lake Mälaren the pilot project seems to have been a Cistercian nunnery church at Sko on lake's northern shore. It was influenced by Danish architecture and probably built c. 1215–20.[8] Sko strongly inspired the Dominicans when they built a monastic church in Sigtuna dedicated to the Virgin. This project probably started in late 1220s, and became a prelude to an intense use of the new building material, not only for building ecclesiastical buildings and fortifications. Thus the Swedish Bjälbo dynasty, which had close relations with Denmark and Germany, erected several grand palatial brick buildings in Östergötland, Västergötland and in the Lake Mälaren region already about the middle of the 13th century.[9]

About the middle of the 13th century and probably at the initiative of Birger Jarl, a brick palace was built in (still pre-urban) Stockholm, south-east of the tall watchtower called Tre kronor. Parts of the foundations of this palace have recently been found. It is probably identical with "the fair house" that according to Duke Eric's Chronicle (Sw. Erikskrönikan) was built by Birger Jarl as part of the new town. It is likely that the first preserved letter where

---

[1] Ljung et al. 2013; cf. Lamm, forthcoming.
[2] Als Hansen 1985, 7.
[3] Nordeide 1983, Abrahamsen and Nordeide 1985, 197.
[4] Rytter 2001.
[5] Knapas 1974, 11, Figure 1; Kuokkanen 1981, 45.
[6] Strucke in Ljung et. al 2013, 25–29.

[7] Kock 1973, Johannsen and Møller 1974.
[8] Redelius 2006.
[9] Lovén 2000.

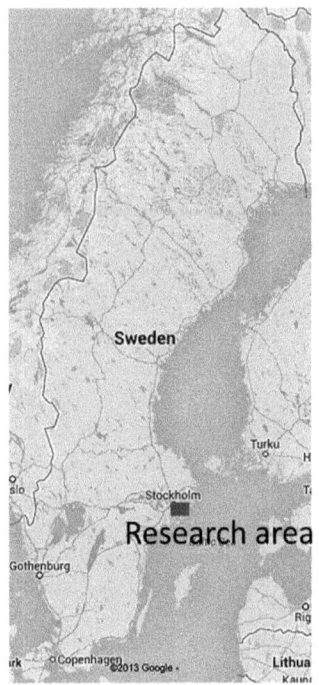

Figure 1. The Research area and the location of the brick kiln/s? RAÄ Boo 22 and 81 (Maps compiled from Google maps by A. Lindahl).

Stockholm is mentioned was written there.[10] In the later 13th century King Magnus Birgersson (Ladulås) pursued several other great building projects including a Franciscan friary on Riddarholmen in Stockholm. King Magnus' interest in favoring brick building is stressed in a letter of donation from 1288 in which he grants the Franciscan friars in Stockholm, together with the Franciscan nuns of St. Claire's nunnery nearby, the use of clay, sand and forest (that is, everything needed for brick-making) at the rural hamlet of Rörstrand, on the northern periphery of modern Stockholm. The donation was confirmed in 1290 when the religious houses were given permission to mine and kiln limestone on Runmarö in the Stockholm archipelago.[11] Clearly the king favored the Franciscan order. Consequently the friary's church became the royal burial church.

The Kiln at Boo

Location and discovery

The site is located on a property owned by J.P. Lamm at Boo on the shore of Baggensfjärden, a bay of the Baltic (Figure 1). In the High Middle Ages the site formed a neck of land at the mouth of a narrow passage, which at that time was the main sea route from the Baltic to Stockholm and Lake Mälaren. Between the house and the seashore there was an odd mound. It looked like a small volcano with a central crater surrounded by gently sloping sides and with lots of smaller and bigger pieces of red brick and white limestone visible on its bramble-covered surface. It was suggested that this was the remains of a building, burnt by Russian troops in connection with an amphibious battle here in 1719. Eventually this interpretation was widened and considering the strategically favorable position at the mouth of the sea route the hypothesis became that the building might have been a watch tower, a brick kiln or maybe a predecessor of Boo manor.[12] As the site was unlisted and not mentioned in any historical documents, it was registered by the Swedish National Heritage Board in its Sites and Monuments Register in which it appeared as #22 in Boo parish under the heading "remnants of a ruined fort".

Excavation

As time passed it became increasingly unsatisfactory not to know what was hidden in the mound next to the house. In 2006 a grant from the Stockholm County Archaeologist enabled excavations at the site. This developed into a research project in collaboration with the archaeology section of the Stockholm County Heritage Society (Sw. Stockholms läns hembygdsförbund), mainly staffed with volunteers from its local branch, and headed by professional field archaeologists. Financial support was provided by several private and public sources.

Soon the presumed watch tower revealed itself as a High Medieval brick kiln (Figures 2–3). Thus Nacka municipality can proudly reckon its industrial history back to the 13th century.[13] At that time brick kilns were usually established either at construction sites for churches, forts, castles and palaces, or, as in our case, at a sea route that enabled

---

[10] Friman and Söderström 2008, 18–19; Söderlund 2010, 752; Regner 2013, 47.
[11] Ankarberg and Nyström 2007, 41.
[12] Lamm 1987, 15.
[13] Sillén 2012, 8–9.

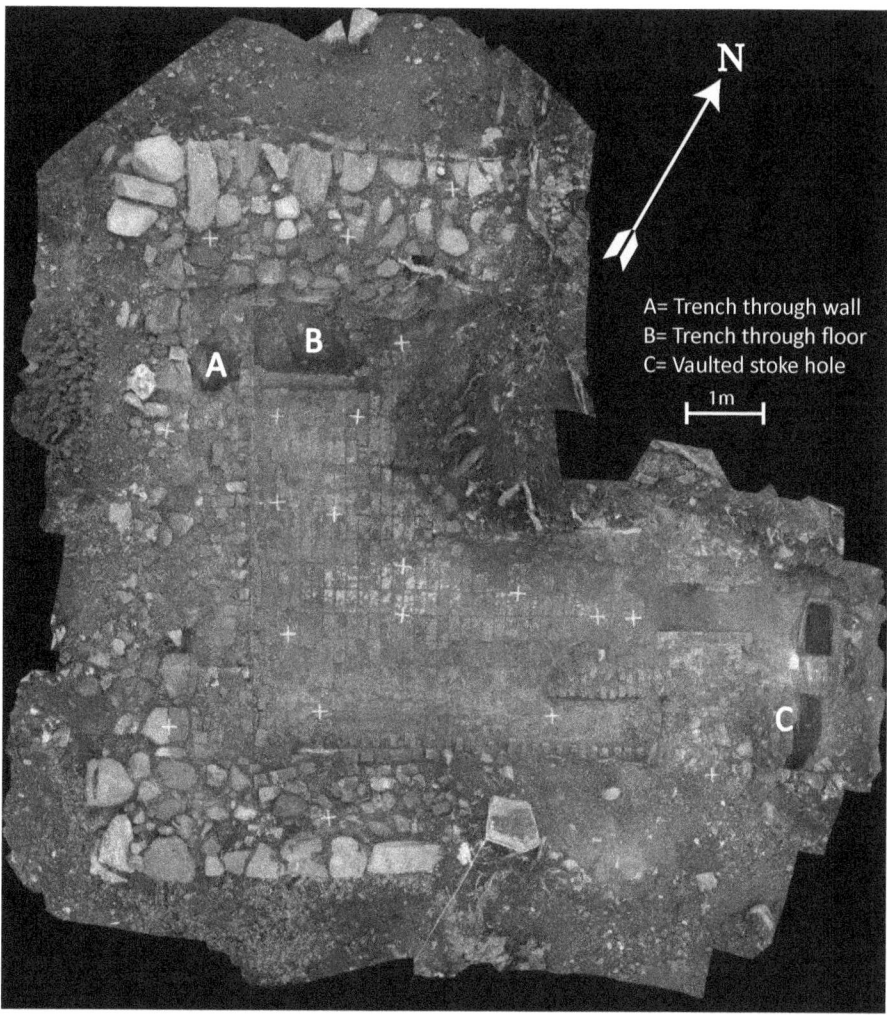

Figure 2. Orthophoto of the kiln. Scale bar 1 m. (Accomplished by J.-Å. Ljung and U. Strucke).

Figure 3. The interior of the kiln seen from SW. In the right hand corner the still intact vault of one of the four stoke pits. The archaeologists are preparing the orthophoto. (Photo by J.P. Lamm)

shipping of their products. Finds of medieval bricks from the bottom of the bay just off the property plot, show that some bricks were dropped during loading. These bricks were undoubtedly being destined for transport to Stockholm. We have also found an area of further High Medieval activity about 50m south-east of the kiln with much brick waste and limestone pieces visible in the ground surface, possibly indicating yet another kiln. It has been registered as # 81 in the Sites and Monuments Register. The excavations attracted great interest. For instance, one of the funding bodies, the Berit Wallenberg Foundation, selected it for an article in their annual report for 2010.[14] The excavation was also presented in a program on Swedish TV.[15]

Results of the excavation

Though there were no economical means to excavate the kiln entirely, we have reached a rather good idea of its chronology and construction history. Based on statistical

---

[14] Tjärnlund 2011.
[15] 15 Produced by Mikael Agaton and shown for the first time in SVT 2 on March 5th 2012.

evaluation of a series of radiocarbon and TL analyses, four stages have been identified between c. 1250 and 1500. However, because of the many transformations the kiln has undergone it has only been possible to get a clear picture of the structure in its final stage.

About ¾ of the total area of the kiln have been uncovered, and trial trenches have been dug outside of the front wall and from this at a right angle down-slope to the 13th century sea level. In 2011 the Department of Geology, Stockholm University, performed diatom analyses[16]. This was continued by an extensive series of XRF analyses of bricks from the kiln by the Laboratory of Ceramic Research at the Department of Geology, Lund University.

From a logistical point of view, the positioning of the kiln at the main route to Stockholm was well chosen. It was easy to supply with firewood, sand and limestone, the latter probably shipped from Runmarö, an island c. 25 km. east. And at the northern edge of the site, a brook delivered fresh water. There was also a good supply of reed that could be used for making mats to protect the rawlings (i.e. unfired raw bricks) from rain.

The kiln was clearly built by expert masons who well knew how to use the topography of the site. They choose a NE–SW depression between two rock outcrops for the foundation of an open-roofed rectangular kiln with thick walls of heavy fieldstones and, probably already from the beginning, a brick floor. Externally it measured 9.05 x 8.35m. As time passed the walls and floor suffered damage from the heat and were mended and rebuilt. The walls therefore acquired a brick-built inner covering in shell-wall technique and new floors laid on top of the old ones. In its earliest stage the kiln had an internal floor area of c. 38m2 and in its latest only c. 26m2. At least three floors have been found. The two latest are well preserved. Counted from the latest floor, up to 15 courses (1.8m) of the rear wall are still standing. However, it was impossible to estimate the original height of the walls.

By the time the kiln was abandoned its floor area had dwindled to c. 5 by 5m. The last front wall was mainly built of brick and had four stoke-pit arches, of which the northeastern most survives intact. At a right angle from the front wall and with the same width as the stoke pits the floor is burnt dark, showing that temporary firing tunnels had been joined to the arches. Between the tunnels the brick floor is unaffected. Thus, judging from the remaining piles of fired brick, it was evident that a channel structure reminiscent of horizontal chimneys had been built probably, with rawlings, before each firing. The channels were oriented NW–SE – in the opposite direction than the prevailing summer winds on site. Between and on top of the channels the kiln had been filled with rawlings, probably up to a height of some meters. Outside the front wall a complicated stratigraphy of material thrown or raked out of the kiln was observed, such as unsuccessfully fired bricks, ashes and powder-burnt limestone. This also formed part of an upwardly growing working-platform from which debris from the firings had been flung towards the shore.

When excavations began the abandoned kiln had a compact fill of brick rubble in which no stratigraphy was discerned.

Production

The first production period of the kiln coincides with the foundation of Stockholm (officially 1252) and the concomitant start of intensive construction. The art of making brick was apparently introduced by monks and the art of using the new building material by Danish or German masons.

During the productive years of the kiln hundreds of thousands of bricks and great quantities of limestone must have been fired and shipped away. The only imaginable market for the building material was nearby Stockholm with its great demand for brick and lime, both from the secular power that built palaces and fortifications and the ecclesiastic ones that erected churches and religious houses.

The kiln made good bricks. However, with the exception of a few hewn and moulded specimens, only ordinary rectangular bricks were made. The measurements correspond to the then prevailing "North European standard" using bricks of tall dimensions (Sw. munktegel), measuring c. 29.5–30cm in length, 13.5–14cm in width and 8–9cm in thickness, weighing about 6 kg.[17]

Industrial logistics

The practical logistics of our brick yard is still largely an open question. For instance it has not been possible to identify with certainty where the clay was quarried. The adjoining beach meadows are a good candidate. Fractured bricks show that the tempering is partly composed of crushed brick (grog). No sand deposits are found nearby, and so the sand needed for tempering must have been shipped some distance, probably from the neighbouring island of Ingarö across Baggensfjärden bay. Great quantities of firewood must also have arrived by boats as well, and all the limestone that was burnt in the kiln. There are no local limestone deposits. The nearest one is the aforementioned one on the island Runmarö, about 25 kilometres east.

In later centuries brick has been dried in special barns, but there are no indications that our brick yard or any of its contemporaries included such structures. Many bricks from Boo have imprints on the upper flat side of dog paws and deer hoofs as well as children's footprints. Therefore, the bricks must have been placed directly on the ground for drying. It must have been protected somehow from rain and too strong sunshine, probably as mentioned by reed or straw mats.

---

[16] Risberg 2013, 36–37, c.f. XRF-analyses on clay by Bjursäter and Helgesson 2012.

[17] C.f. Johannsen and Møller 1974.

Another basic question that cannot be answered with certainty is the capacity of the kiln. In a contemporary kiln at Kungahälla in western Sweden near Gothenburgh the production per firing has been calculated at 12,000 bricks placed on the edge and piled in 20–22 courses. Six firings a year – only during the summertime when the clay would not freeze – would correspond to an annual output of c. 73,000 bricks. But only about 60% of those are believed to have achieved acceptable quality, i.e. c. 43,800. The staff of workmen for the direct production has been estimated to 6–10 persons.[18]

Historical setting

The kiln at Boo does not seem to be mentioned in any written sources. It was also unknown to the locals up until the excavations began and no tradition about it survived. The kiln may have been founded early in the 13th century and can thus be coeval with the earliest brick buildings in the Lake Mälaren area. But it is more likely that it was founded around the middle of the century. We are unlikely to ever learn at whose initiative it was built. Certainly it must have been someone with very good connections and economic resources. It may have been the king himself or his commander-in-chief, the Lord High Constable (Sw. marsk). A letter of donation given in 1282 by King Magnus shows that the king owned land on Hargsö (lat. Arxö), the ancient name of the land-locked former island on which the kiln was built. He donated it to one of his favorites, Tideman Fries, as a reward for services given.[19] Perhaps a substantial part of the value of the gift was a profitable brick kiln? In 1285 Fries also acquired a part of Arxö that has belonged to Karl Gustavsson, the marsk. In 1323 Fries' widow sold the property to the archbishop.

Bo/Boo was originally a noun meaning "manorial farm acting as administrative centre of a land estate", and here this classificatory word seems to have superseded an earlier name for the manor, Hargsö[20]. The estate remained property of the church until the Reformation, and it is likely that in the 14th century the archbishop leased it to a famous merchant, Conradus de Arxö (obit. c. 1340)[21] and his sons, who probably run the kiln as one of their many businesses. A combination of TL and radiocarbon dates suggests that the last firing of the kiln took place between 1380 and 1500. The only datable artifact found in the kiln is part of a tripod pot from c. 1590, but at that time the kiln had probably been abandoned for about a century.

The Kiln at Boo compared to Danish parallels

In connection with the investigation of a fortification at Søborg on Zealand a brick kiln was excavated. It was TL dated to 1160±50 years. Thus it belongs to the earliest scientifically dated kilns in Denmark. A better-documented one, however, is a medieval brick yard at Bistrup in Roskilde.[22] There five kilns were excavated in 1975–6. Judging from the bricks found, the earliest kiln is believed to have been founded c. 1250. Stratigraphically it was succeeded by a kiln which remained in use until about 1550.

As to size and construction both kilns are similar to the one at Boo. According to Als Hansen and Aaman Sørensen they both correspond to all other kilns they know about from medieval and modern times. Briefly, they describe these brick kilns as masonry boxes without a lid and with firing holes in the front wall. Building material and size may differ, and floor and firing channels vary, but the basic design hardly changed from the Middle Ages until the 20th century.

The authors conclude that the kilns always have a right-angle configuration, often approaching a square. Like at Bistrup they were very often dug into a hillside. Generally the rear wall is much better preserved than the front wall. The height varied with the size of the kiln, but according to a Danish rule of general application a kiln should contain at least 20 courses of brick for a successful firing. This means that the minimum height lies somewhere between 2.5 and 3.0m. The walls generally have a shell of burnt brick that has been laid against a rear of stone, green brick or clay. For mortar a more or less sand-mixed and sometimes lime-mixed clay mortar was used. The kilns may differ much in size. The depth – i.e. the distance between front and rear wall – can vary between c. 3m and 7m and the length between the side walls between 3m and over 9m – this depending on the number of firing holes. These holes in the front wall are vaulted, normally between 60 and 80cm wide and up to 1m high.

The floor in its simplest execution is quite flat and may consist of clay or brick. Generally, low benches have been built along the side walls and between the firing channels and the rear wall. On the benches or on the flat floor between the firing holes the rawlings were piled, forming temporary firing channels. The rawlings were piled at some centimeters' distance from each other to allow hot gasses from the firing channels to penetrate among the rawlings. The kiln, as mentioned, had no roof. When it was almost full the rawlings were piled closer to close it up. Finally the top surface was covered with waste brick, clay and turf. At certain distances vent holes were left. In some cases permanent firing channels were built. They were composed of arches between which hot gas could ascend.

Kiln 1 at Bistrup was about the same size as the Boo kiln in its earlier stage. It was calculated to have the capacity to fire 20,000 – 30,000 bricks at a time, and the largest Danish kilns allowed c. 40,000. Misfirings were common

---

[18] Rytter 2001, 140–3.
[19] Jansson 1943, 19–21. According to Nils Ahnlund (1953, 126, 146) Fries probably belonged to the grand-bourgeois family Fries in Lübeck and was one of the first mayors of Stockholm. In contemporary documents he is called "Dominus".
[20] Ringstedt et.al. 2007, 15.
[21] Conradus de Arxö belonged to the leading grand-bourgeois in Stockholm and obviously had got his cognomen from Hargsö (Ahnlund 1953, 160).

[22] Als Hansen and Aaman Sørensen 1988.

and could have many causes, e.g. poor drying, faulty piling or bad covering. Thus great quantities of faulty burnt brick accumulated around the kilns. No convincing traces of drying barns have been found in medieval Danish brick yards.

hXRF-analyses of the bricks from Boo

Material

The material encompassed in the hXRF-analysis consists of 75 samples. They were all selected from whole bricks. The size of the bricks varies slightly (length 280 – 310mm, width 130 – 147mm and thickness 85 – 105mm), however, this variation may be considered normal for handmade bricks and most likely made in different moulds and exposed to slightly different firing temperatures and thus display differences in shrinkage. Unfortunately none of the bricks can be related to any specific time or sequence in usage of the kiln.

Aims and objectives

The main objective of the hXRF-analysis was to test the variation in chemical composition on a material which in all likelihood was made locally from (presumably) the same raw clay over a limited period of time – in this case c. 200 years.

The aims of the study are to:

- determine what may be considered as a normal variation within a sample from this type of context.
- see if it is possible to distinguish any groups within the sample.
- explain anomalies in the sampled material.
- exemplify methods on how to describe qualitative differences in a ceramic material.

Method

X-ray fluorescence (XRF) is a non-destructive method for chemical analysis in order to define the elemental composition of a test sample. For the analyses of the bricks from Boo a Thermo Scientific portable (handheld) XRF analyzer (hXRF), Niton XL3t 970 GOLDD+ has been used. The method has highly accurate determinations for major elements[23]. A limitation in the XRF-analysis with this equipment is that elements in the range from Sodium (Na) and lighter cannot be detected. The elements that are analysed in this investigation are: Mg, Al, Si, P, S, Cl, K, Ca, Ti, V, Cr, Mn, Fe, Co, Ni, Cu, Zn, As, Se, Rb, Sr, Y, Zr, Nb, Mo, Pd, Ag, Cd, Sn, Sb, Ba, W, Au, Pb, Bi, Th and U.

Clay, and by default ceramics, consist mainly of the elements Silicon (Si) and Aluminium (Al) in the oxide states SiO_2 and Al_2O_3 as well as quite a large portion of Iron (Fe_2O_2/Fe_2O_3), Calcium (CaO) and Potassium (K_2O). All these are normally found in percentage amounts. Other elements are only detected as parts per million (ppm).

All the analyses have been performed on a polished breakage of the samples. With the Niton XL3t it is possible to apply either a 3mm or 8mm spot radius. In order to get as good representation of the analysed object as possible the 8mm spot was used.

The element detection is also depending on the duration of the analysis. This means that a higher accuracy in the quantitative analysis is achieved if the object is analysed for a longer time. This is especially true for the lighter elements in the analysis such as Mg, Al, Si, P, S, Cl, K, Ca. For this reason the measuring time is set to 6 minutes.

Results

Several of the elements that partake in the analysis are missing or present in amounts under the detection limit in the brick material from Boo. In the comparison between the samples the range of elements is for this reason limited to: Mg, Al, Si, P, S, K, Ca, Ti, V, Cr, Mn, Fe, Cu, Zn, As, Rb, Sr, Y, Zr, Nb, Ba, Pb, Th and U.

As an initial comparison the two major elements Si and Al are compared in a two-dimensional plot (Figure 4). What is very evident from this plot is that there is a concentration that contains most of the samples. There are a few outliers that are clearly not part of this concentration such as samples 5, 10, 46, 55, 67, 70, 71 and 72.

The two elements Potassium and Rubidium normally display good correlation in a plot (Figure 5). This is also true with the samples from Boo. With the exception of sample 10 they all form a nice correlation. There are a few other samples e.g. 19, 46, 55 and 70 that deviate slightly from the main cluster of samples.

Further comparisons of two-dimensional plots between the other elements display similar pictures with the bulk of the material in a concentration and a few samples as outliers. The number of outliers as well as the identity of the samples may vary from plot to plot, but the compiled result of this type of comparison indicate that samples 5, 7, 10, 19, 46, 55, 67, 68, 70, 71 and 72 are made of a clays of a different composition than the bulk of the material.

Apart from the bulk of the material there is only one set of 9 samples (samples 3, 10, 24, 31, 36, 40, 46, 53 and 80) that can be said to form a group (Figure 6). The main element that differs from the bulk of the material is Calcium (Ca). Depending on the amount of Ca the group may be further divided into three sub groups A (24, 31, 40), B (3, 53, 80) and C (36, 46, 10).

To get a more comprehensive statistical evaluation involving all the elements in the XRF-analysis a multivariate

---

[23] Helfert and Böhme 2010; Helfert et al. 2011; Papmehl-Dufay et al. 2013.

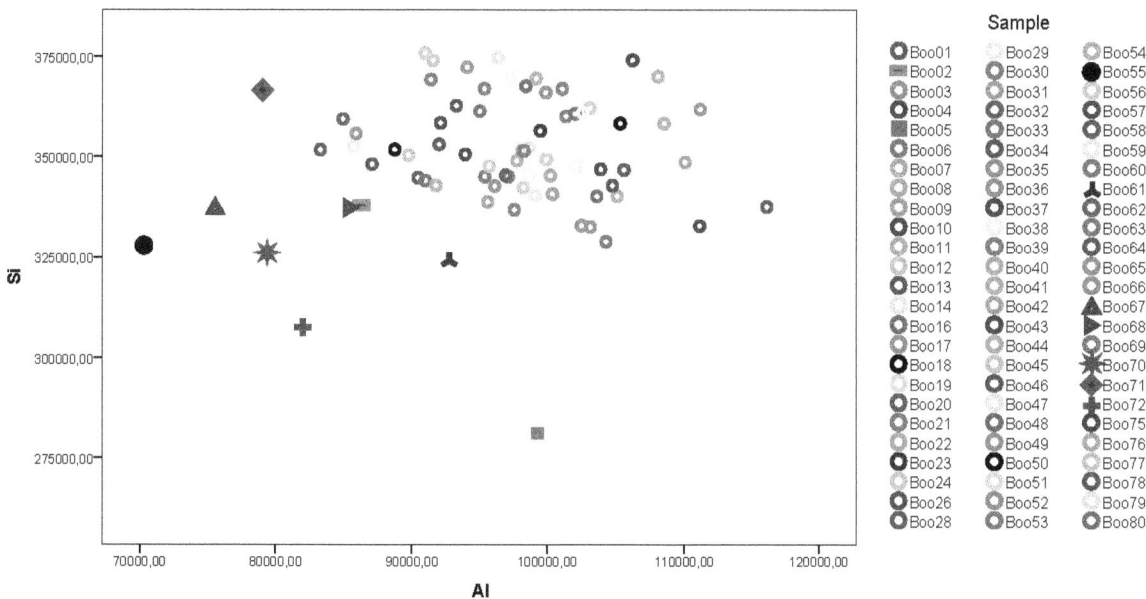

Figure 4. Bivariate scatterplot between the elements Silicon (Si) and Aluminium (Al). Most samples are concentrated in a cluster. The samples that clearly differ from the main material are sample 5 with low Silicon content and samples 55, 67, 70, 71 and 72 with low Aliminum content. (Chart by A. Lindahl)

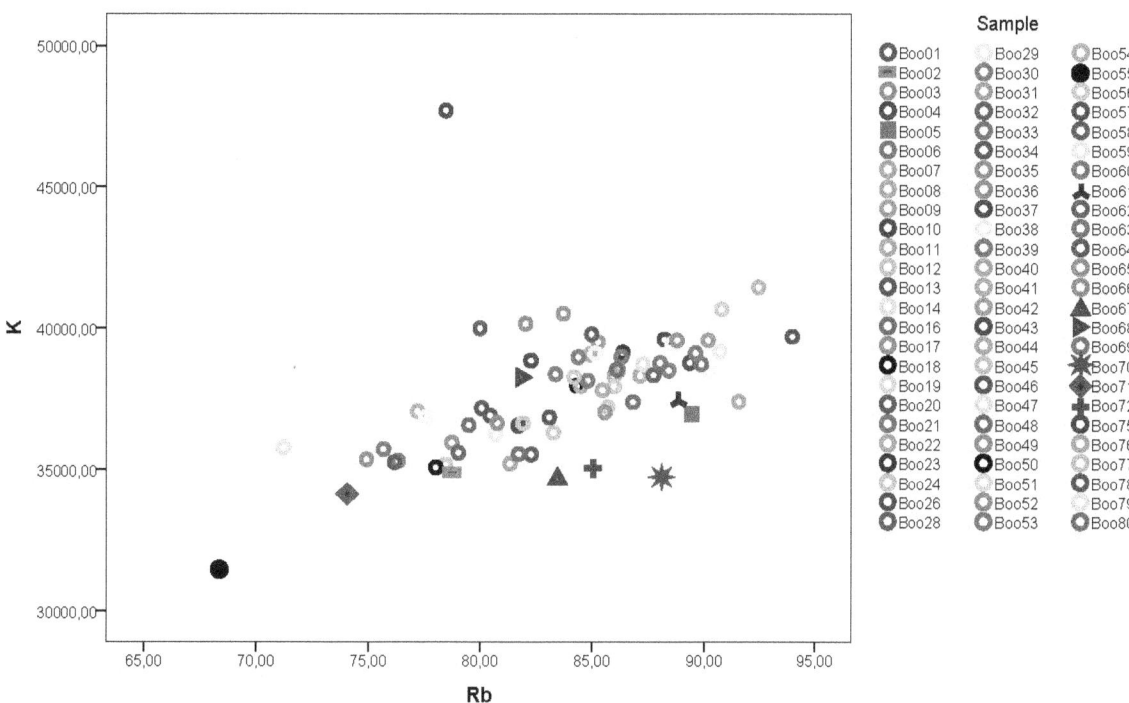

Figure 5. Bivariate scatterplot between the elements Potassium (K) and Rubidium (Rb). With one exception - sample 10 with very high content of Potassium - all samples are nicely correlated. Also sample 55 falls outside the main cluster of samples. (Chart by A. Lindahl)

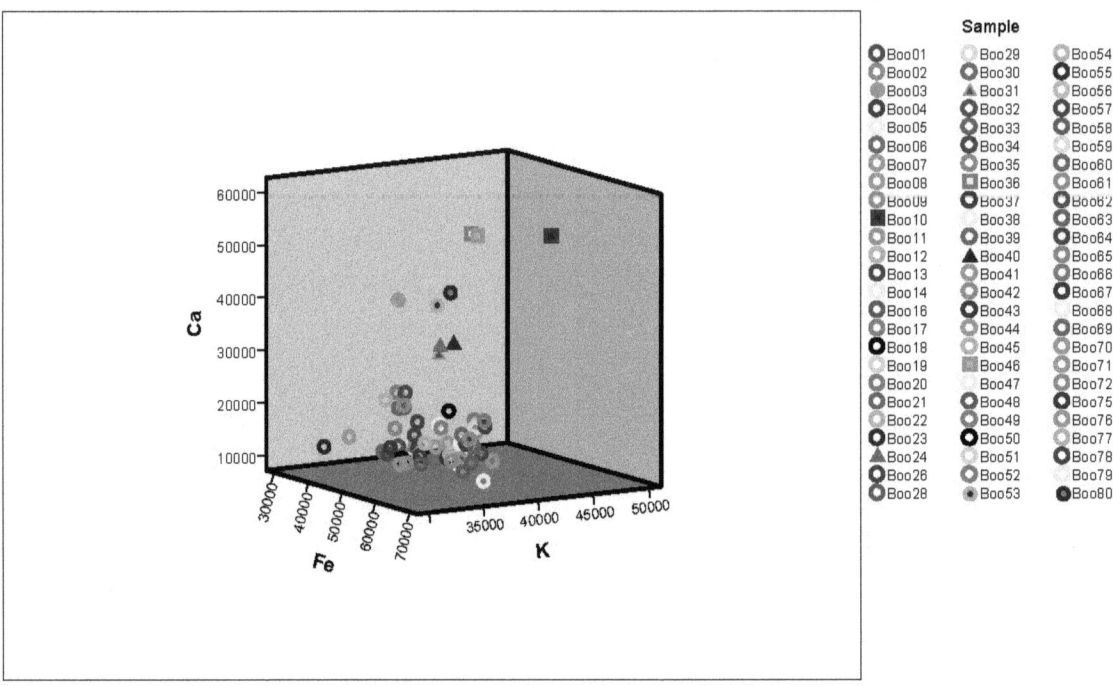

Figure 6. 3-dimensional scatterplot between the elements Calcium (Ca), Iron (Fe) and Potassium (K). The samples that stand out have a high Calcium content and consist of three groups with three samples in each group. Sample 10 also contain a high Potassium content. (c.f. Figure 5) (Chart by A. Lindahl)

analyses, Principal Component Analyses (PCA)[24], was carried out using the statistical software package IBM SPSS Statistics version 21. More than 98% of the total variance was explained in the three first components. These factor scores from these components were then plotted against each other in a three-dimensional plot (Figure 7). The trend in distribution that could be observed in the previous diagrams is now much clearer. The bulk of the material forms a very neat cluster. The samples with high Ca content clearly stand out as well as samples 5, 55, 67, 70 and 72. Some samples that were previously noted as deviating from the main material e.g. samples 7, 19, 68 and 71 are now part of this although 68 and 71 together with two "new" samples (2 and 61) are on the very edge.

Calcium seems to have a very strong impact on the analysis and to determine the impact of this element a second PCA-analysis excluding calcium was performed. In this new comparison 97% of the total variance was explained in the two first factors. In the plot between these two factors all the Ca rich samples now are part of the main material (Figure 8). Samples 5, 55, 67, 70, 71 and 72 are clearly different than the rest of the material and samples 2, 61 and 68 are slightly to the side of the main material.

The scatter plots are a very good to display data and they are also form a good basis to interpret how the different samples are connected and grouped. However, this interpretation is subjective and in order to get a more objective interpretation of the PCA-analyses the factor scores were tested by means of a hierarchical clustering[25]. The test including calcium shows the calcium rich samples as different compared to the main material, but what is more evident is that sample 5 has no similarity with any other sample. Samples 55, 67, 70 and 72 also have a very different composition than the rest of the material (Figure 9).

Another test in which calcium is excluded from the PCA-analysis displays a similar distribution (Figure 10). Samples 5, 55 and 72 are different from one another and the rest of the material. Another set of samples that branches out from the bulk material consists of one single sample (71) and two groups, one consists of samples 2, 67, 68 and 70 the other group which is very homogenous encompass samples 1, 13, 22, 29, 32, 34, 50, 66, 76 and 78.

Conclusions

The excavations at Boo revealed one of the earliest brick kilns in Sweden. TL- and 14C- datings conclude that the kiln was in use approximately 200 years, from 1250 to 1450.

Since no clear stratigraphy could be observed among the bricks we must assume that the sampled material represents the whole lifespan of the kiln. The favourable location of the brick-yard at the entrance to the narrowest passage of the fairway to Stockholm implies that its manager must have had the best possibilities to continuously keep himself informed about matters important for his craft and business

---

[24] Hotelling 1933.

[25] Hastie, Tibshirani and Friedman 2009.

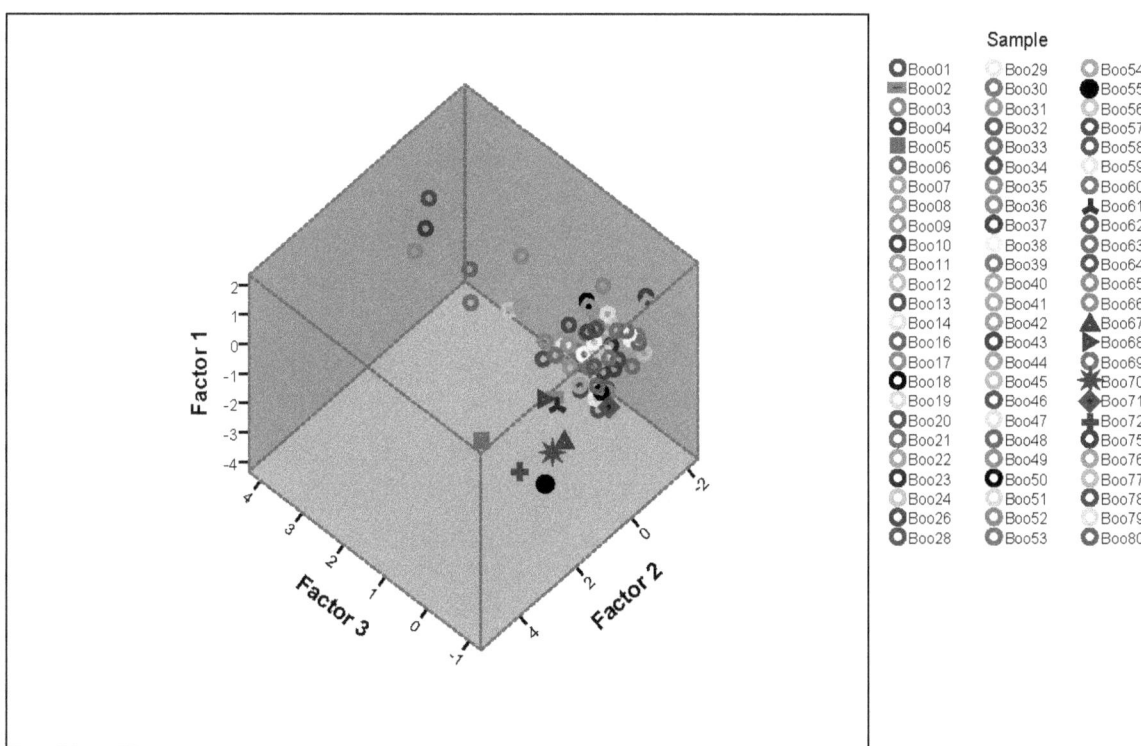

Figure 7. 3-dimensional scatterplot between the three first factors of a Principal Component Analysis (PCA). The used elements are Mg, Al, Si, P, S, K, Ca, Ti, V, Cr, Mn, Fe, Cu, Zn, As, Rb, Sr, Y, Zr, Nb, Ba, Pb and Th. The Calcium rich samples stand out in the plot as do the samples 5, 55, 67, 70, 72 and to some extent samples 2, 61 and 68. (Chart by A. Lindahl)

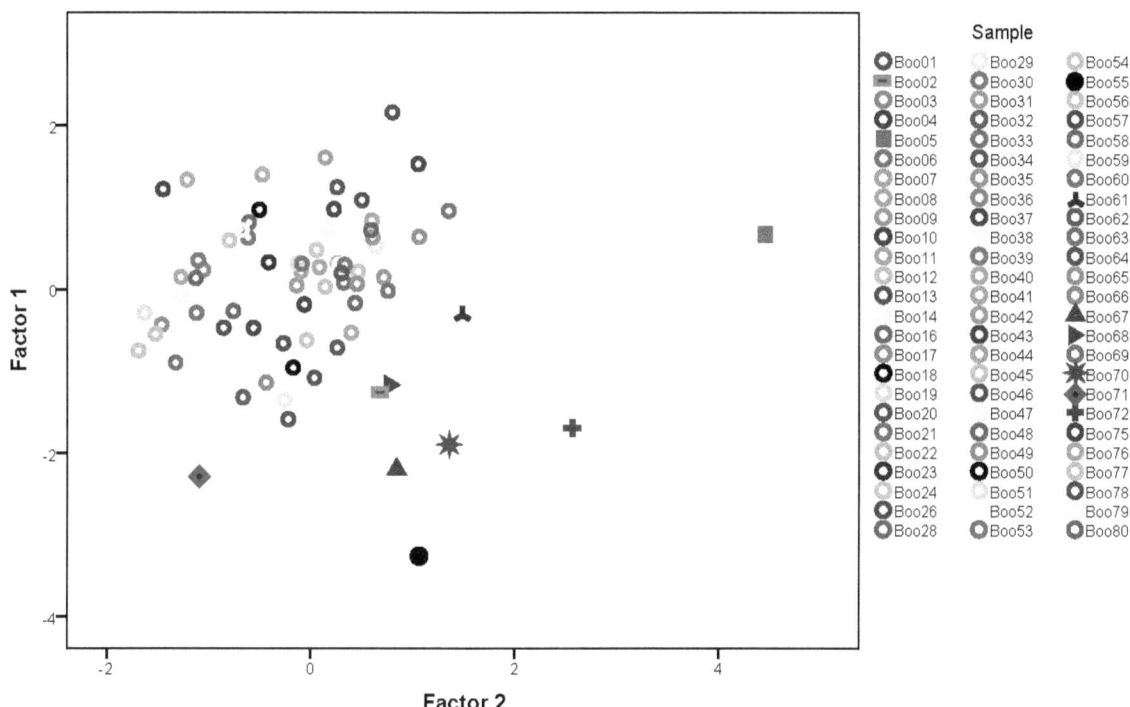

Figure 8. Bivariate scatterplot between the two first factors of a Principal Component Analysis (PCA). The used elements are Mg, Al, Si, P, S, K, Ca, Ti, V, Cr, Mn, Fe, Cu, Zn, As, Rb, Sr, Y, Zr, Nb, Ba, Pb and Th. The vast majority of the samples form a cluster, the samples that clearly differ are samples 5, 55, 67, 70, 71 and 72 and to some extent samples 2, 61 and 68. (Chart by A. Lindahl)

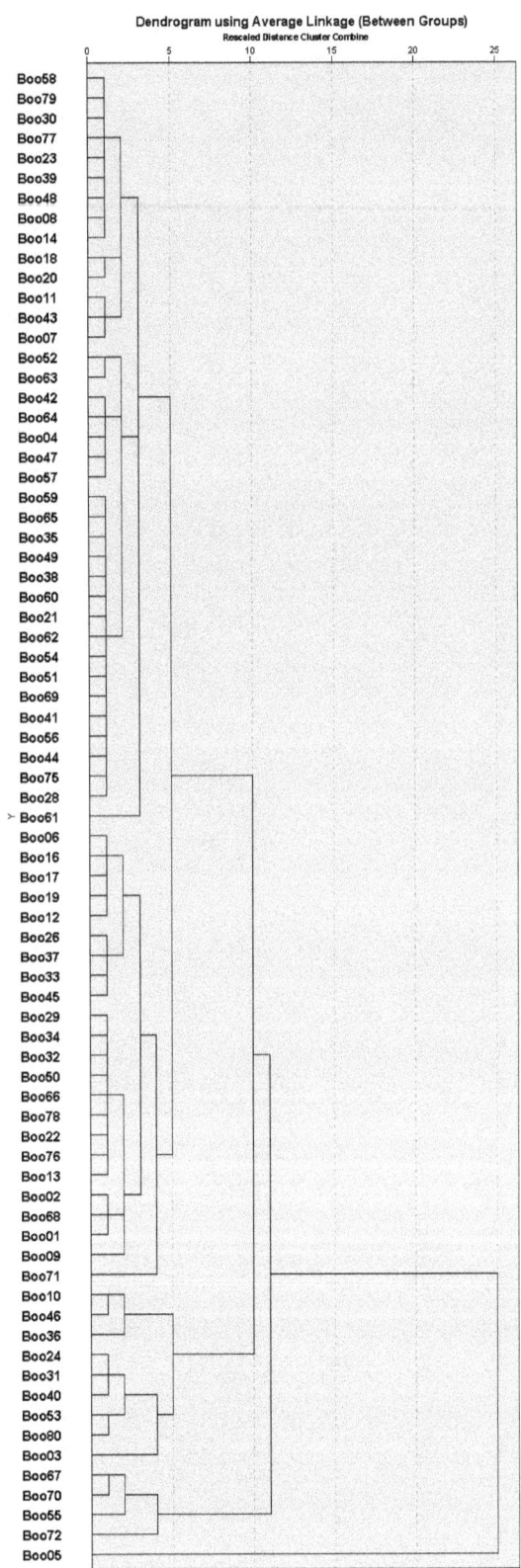
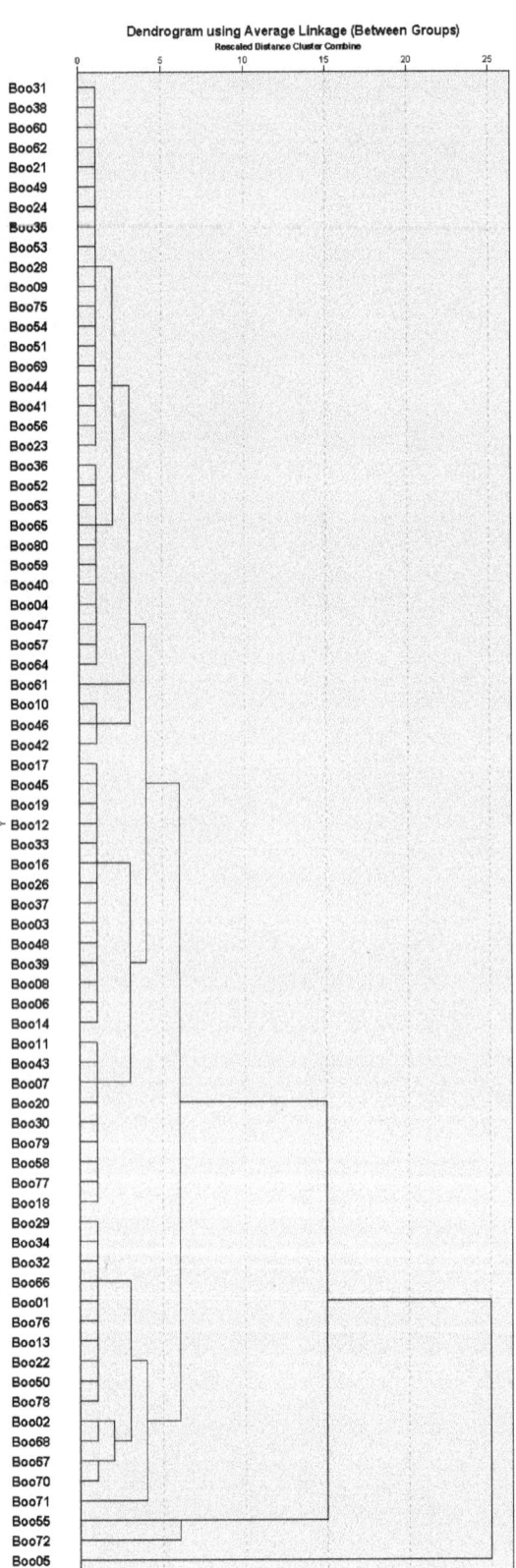

Figure 9. Dendrogram of a Hierarchical Cluster Analysis using the three first factors of a Principal Component Analysis (PCA). The used elements are Mg, Al, Si, P, S, K, Ca, Ti, V, Cr, Mn, Fe, Cu, Zn, As, Rb, Sr, Y, Zr, Nb, Ba, Pb and Th. The majority of the samples are part of two larger concentrations. The Calcium rich samples form a group of their own. The samples 55, 67, 70 and 72 form a third cluster. Sample 5 is singled out as having no connection to any other sample. (Chart by A. Lindahl)

Figure 10. Dendrogram of a Hierarchical Cluster Analysis using the two first factors of a Principal Component Analysis (PCA). Ca is snow excluded in the analysis and the used elements are Mg, Al, Si, P, S, K, Ti, V, Cr, Mn, Fe, Cu, Zn, As, Rb, Sr, Y, Zr, Nb, Ba, Pb and Th. The majority of the samples are part of three larger concentrations. Three samples have a completely different chemical composition than the rest of the material, samples 55 and 72 which are loosely connected and sample 5 which is singled out by having no connection to any other sample. (Chart by A. Lindahl)

from passers by staying at the site or in the nearby harbour of Boo.[26] History tells that in the Medieval period the foreign contacts mainly were with Denmark and Germany.

With the exception of a few samples the bricks from Boo display a very homogenous group according to the XRF-analysis. There is a group of 9 samples that display a considerably higher content of calcium than the rest of the material. This anomaly is most likely a result of contamination since the kiln evidently had been used for burning lime.

There are different means to interpret the data from the XRF-analysis, however, they all show very small variations. By studying the scatter plots of two elements at a time and compiling the results there are possibly 10–15 samples that differ more or less from the bulk of the material. However, this is a very time consuming as well as subjective way to interpret the data. When using statistical methods such as PCA (Principal Components Analysis) the understanding of the material may be diminished, but on the other hand a more unbiased interpretation is achieved. In the Boo material fewer samples fall outside the bulk material in this way. There are a number of ways to present the data from the XRF-analysis. The easiest is to use scatter plots with the actual data from the analysis, either as two- or three dimensional plots (Figures 4–6). This means to display data can also be used for the result of the PCA (Figures 7–8). In this case, with a limited number of samples (e.g. less than 100), it is possible to get a view of the material by using the result of the PCA in a dendrogram (Figures 9–10.) As is evident from the analyses especially five of the samples (5, 55, 67, 70 and 72) stand out and display a completely different element composition than the rest of the material. However, with the exception of the five samples just mentioned, it is likely that the clay for making bricks at Boo was quarried from the same deposit and in all likelihood was local.

The future

By the end of the 2010 fieldwork c. 2/3 of the kiln's plan had been uncovered. The structure made a stunning impression with its well preserved walls and almost intact floor. This presented us with a dilemma – what to do with the ruin after the excavation. We were asked to try to secure it for the future as a preserved visiting object. However we realized that this was unrealistic, not least for reasons of Swedish climate with its heavy snows and annual freeze-thaw cycle. Thus we decided to refill the ruin with our spoil dumps. This was done after having insulated the floor and inner walls with styrofoam blocks. Part of the outer wall in the south-west corner was considered strong enough to leave it visible as a teaching aid.

In 2011 a trial trench was dug from the front wall down the slope of the hill in a northerly direction. In the trench, which passed through c. 2m thick deposits of debris from the kiln, diatom samples were taken by geologists. They showed the sea level when the first kiln was built, and the results can hopefully contribute to the discussion on the progress of shoreline displacement in the region.

The kiln mound is now back to its general former appearance. There are good possibilities for future archaeologists to continue and reexamine our investigation. Hopefully they can answer many questions that were beyond our resources. By detailed study of sections through the walls and the floors they might find out how the kiln was constructed in its earliest stage and collect further information about how it changed over time.. A more urgent task is a trial excavation of the medieval activity area that we have found on the adjoining plot south of the kiln. As mentioned, it probably conceals another kiln for burning brick and limestone.

English revised by Martin Rundkvist.

Acknowledgements

The hXRF-analyses were made possible due to generous financial support from the Berit Wallenberg foundation.

References

Als Hansen, B. and Aaman Sørensen, M. 1980. Bistrup Teglværk- Arkæologiske undersøgelser af en middelalderlig teglværkstomt ved Roskilde. Hikuin 6, 221–64.

Abrahamsen, N. and Nordeide, S.W. 1985. Magnetiske undersøgelser af en middelalderteglovn fra Tønsberg, Norge. Kuml, 187–189.

Ankarberg, C. H. and Nyström, B. 2007. Rörstrand i Stockholm. Tegelbruk, fajansmanufaktur och keramisk storindustri 1270–1926. Stockholm.

Ahnlund, N. 1953. Stockholms historia före Gustav Vasa. Stockholm.

Bjursäter, S. and Helgesson, T. 2012. En jämförelse av XRF-mätningar på medeltida tegel och nutida leror från Saltsjö Boo, Nacka kommun. Studentarbete. Kvartärgeologiska institutionen. Stockholms universitet.

Cederlund, C.-O. 1989. Sjövägarna vid svenska östersjökusten på 1200-talet. Del 2. Marinarkeologisk tidskrift.

Friman, H. and Söderström, G. 2008. Stockholm: en historia i kartor och bilder. Stockholm.

Frödin, O. 1919. Från det medeltida Alvastra. Fornvännen 14, 43–68.

Hastie, T., Tibshirani, R. and Friedman, J. 2009. 14.3.12 Hierarchical clustering. The Elements of Statistical Learning (2nd ed.). New York: Springer, 520–528.

Helfert, M. and Böhme, D. 2010. Herkunftbestimmung von römischer Keramik mittels protabler energiedispersiver Röntgenfluoreszenzanalyse (P-ED-RFA) – Erste Ergebnisse einer anwendungsbezogenen Teststudie. In: B. Rammingar, O. Stilborg (eds), Naturwissenschaftliche Analysen vor- und frühgeschichtlicher Keramik I.

---

[26] Cederlund 1989.

Universitätsforschungen zur prähistorischen Archäologie 176, Bonn.

Helfert, M., Mecking, O., Lang, F. and von Kaenel, H.-M. 2011. Neue Perspektiven für die Keramikanalytik. Zur Evaluation der portablen energiedispersiven Röntgenfluoreszenzanalyse (P-ED-RFA) als neues Verfahren für die geochemische Analyse von Keramik in der Archäologie. Frankfurter elektronische Rundschau zur Altertumskunde 14, 1–30.

Hotelling, H. 1933. Analysis of a complex of statistical variables into principal components. Journal of Educational Psychology, 24, 417–441 and 498–520.

Jansson, E. A. 1946. Boo sockens historia. Stockholm.

Johannsen, H. and Møller, E. 1974. Tegl. In: Kulturhistoriskt lexikon för nordisk medeltid, Vol. 18, Sp.149–158. Malmö.

Knapas, R. 1974. Tiilen historiaa. Tiili 1974/4: 10–17.

Kuokkanen, R. 1981. Tiilen lyönti ja käyttö ristiretkiajalta 1850-luvulle, I osa. In: Suomen tiiliteollisuuden historia, 9-185, Suomen tiiliteollisuusliitto r.y and Tiilikeskus Oy, Helsinki.

Lamm, J. P. 1987. En 1500-talsteckning av Jacob Bagges Boo. Värmdö skeppslags fornminnesförenings årsskrift, 11–15.

Lamm, J.P. Forthcoming. Antikvariskt varjehanda från Boo 1980–2013. In: Värmdö skeppslags fornminnesförenings årsskrift 2012–2013 (richly illustrated with photos of the kiln).

Ljung et. al. Ljung, J.-Å.; Strucke, U.; Vinberg, A.; Risberg, J. and Lamm, J.P. 2013. Tegelugnen i Boo. Medeltida tegeltillverkning i större skala Uppland, Boo socken… Riksantikvarieämbetet. UV rapport 2013:27. Stockholm.

Lovén, C. 2000. Folkungapalatsen och deras efterföljare. In: B.I. Johansson and C. Lovén (Red.) Byggnader och betydelser. En antologi om arkitektur, Stockholm, 43–57.

Nordeide, S.W. 1983. Tegl i Tønsberg i middelalderen - Produksjon og produksjonsforhold. Avhandling for magistergraden i nordisk arkeologi. Universitetet i Oslo.

Papmehl-Dufay, L., Stilborg, O., Lindahl, A. & Isaksson, S. (2013). For everyday use and special occasions: A multi-analythical study of pottery from two Early Neolithic Funnel Beaker (TRB) sites on the island of Öland, SE Sweden. In Ramminger, B., Stilborg, O. & Helfert, M. (eds.) Naturwissenschaftliche Analysen vor- und frühgeschichtlicher Keramik III. Universitäsforschungen zur prähistorischen archäologie, Band 238. Bonn, 123–152.

Redelius, H. 2006. Sigtuna och Sko en tid ett rum. Uppsala.

Regner, E. 2013. Det medeltida Stockholm. En arkeologisk guidebok. Lund.

Ringstedt et. al.2007. Ringstedt, N., Rahmqvist, S. and Lamm, J.P. Märklig fornlämning i Saltsjö Boo. Ledungen 1/2007, 15–17.

Risberg, J. 2013. Diatoméanalyser av två sekvenser från tegelugnen i Saltsjö Boo. In:Ljung et.al.2013, 36–38.

Rytter, J. 2001. Teglovnen i Kongehelle In: H. Andersson, K. Carlsson and M Vretemark (eds) Kungahälla. Problem och forskning kring stadens äldsta historia, 134–47. Uddevalla.

Sillén, G. 2012. Kvarnhjul och fabriksskorstenar. Nackas industriarv. Nacka, 8–9.

Söderlund, K. 2010 Stockholm's fortifications in medieval and early modern times. Lübecker Kolloquium zur Stadtarchäologie im Hallserum VII. Die Befestigungen. Hrsg. Für den Bereich Archäologie und Denkmalpflege der Hansestadt Lübeck von Manfred Gläser. Lübeck.

Thordeman, B. 1920. Alsnö Hus. Ett svenskt medeltidspalats i sitt konsthistoriska sammanhang. Stockholm.

Tjärnlund, J. 2011. Villatomten gömde medeltida tegelugn. In: Wallenbergsstiftelserna. 2010 års verksamhet, 11. Stockholm.

UNFIRED BRICKS USED FOR A MEDIEVAL OVEN IN TURKU, FINLAND

Tanja Ratilainen

Abstract: A foundation made of unfired bricks was discovered in the project Early Phases of Turku, which was conducted by the Museum Centre of Turku in 2005–2007. The structure measuring 2.2x2.2m was interpreted as an oven and it was dated to the 1320s based on stratigraphy, radiocarbon dates and dendrochronological results. In this paper, the peculiar oven is presented in detail and its function and purpose is discussed, since nothing exactly the same is so far known from Europe. It is probable that the raw bricks were left over or were the by-products of a large building project going on in the town in the beginning of 14th century and that they were simply applied to the oven that needed to be built.

Keywords: Adobe, medieval, oven, town archaeology, Turku, unfired bricks, raw bricks

Introduction

The town of Turku, located on the southwest coast of modern day Finland, is one of the six towns founded in the eastern part of medieval Sweden (Figure 1). In the Middle Ages (c. 1150–1520)[1] Turku was the most important of them, being the administrative center of the church as well as the center of trade. At the end of the medieval period, the population of Turku was probably about 1500.[2]

In 2004, Turku celebrated its 775th anniversary[3] and, to honor that marker, the town decided to grant funds to carry out excavations to determine how old the town really was. Thus, the project Early Phases of Turku was conducted by the Museum Centre of Turku in 2005–2007. It was the first research excavation in the area outlined by the Great Market, medieval Hämeenkatu street, the cathedral and the river Aura with the possibility of studying all the layers down to the natural clay.[4] With the three excavation areas around the cathedral the oldest cultural layers were expected to be found. (See Figure 1) It was discovered that the square called Horse Market (Area 1), which was presumed to date to the 13th century,[5] was founded there only after the mid-15th century. It was preceded by ordinary town plots with timber buildings. Church Street (Area 2), one of the most important medieval streets, leading from the Great Market to the Cathedral,[6] was first founded only at the beginning of the 14th century, along with the first structures between the Cathedral and the river Aura (Area 3)[7] (Figure 1). In all the areas, under the town layers, plough marks and a cultivation layer radiocarbon dated to the 13th century were discovered. Therefore, it seems that the Cathedral hill called Unikankare was surrounded by fields in the 13th century and the first urban settlement emerged there only in the beginning of 14th century.[8] However, one of the oldest and most striking findings of the excavations was a rectangular foundation made of unfired bricks mortared together with lime. The structure was found in Area 2 in the corner of the oldest wooden building of the site.[9]

To build with unfired bricks is an ancient technique developed in the Near East, dating back to at least 8000 BC. Moulds started to be used during the 6th and 5th millennia BC. The invention of firing the bricks, which increased the durability of the building material, was first introduced c. 5000–4500 BC, but was more widely adopted around 3000 BC.[10] Later the Romans spread their (fired) brick building art around Europe. It was probably in Lombardy, in Northern Italy, where the medieval brick production and brick building skills developed. The first brick buildings were erected in modern-day northern Germany, Denmark and southern Sweden in the second half of the 12th century.[11] In central Sweden, the use of brick was introduced by the mid-13th century.[12] On the other side of the Baltic Sea, in Estonia, the earliest traces date back to the turn of the 13th and 14th centuries and, in Latvia, brick became familiar during the 13th century.[13] In Finland, masonry skills, including brick, arrived at the

---

[1] E.g. Törnblom 1993; Virrankoski 2009. Prior to that the European Early and the beginning of the High Middle Ages i.e. 5th–mid–12th centuries (Graham-Campbell 2007, 17) is still the Iron Age in Finland. (e.g. Hiekkanen 2002, 488; Siiriäinen 1987, 27; Virrankoski 2009, 38–39. See also Helle 2003, 5–6)
[2] Hiekkanen, 1997, 377–378; Törnblom 1993, 377–381.
[3] Counting from year 1229 when Pope Gregory IX ordered the bishop's see to be transferred to a more suitable place. (REA 1, available also online: http://extranet.narc.fi/DF/detail.php?id=72; e.g. Gardberg 1971, 150; Pihlman and Kostet 1986, 18.
[4] Pihlman 2007a, 15; 2007b, 86; 2010, 21–22.
[5] Gardberg 1969, 30–31; 1971, 265–273; Pihlman and Kostet 1986, 63.
[6] See e.g. Ruuth 1909, 64; Pihlman 2007b, 88; Seppänen 2012, 909.
[7] About the first settlers of the town, see: Gardberg 1971, 171–175; Pihlman 2007b, 86.

[8] Pihlman 2007a, 15; 2007b, 88–89; 2010, 21–22; Ratilainen 2007, 23; 2010, 45–55, Saloranta 2007, 31; 2010, 72–73. About the finds see e.g. Majantie 2007, 32–49. The latest overall picture on the development of the town in the Middle Ages: Seppänen 2012 and in English: Seppänen 2009, 240–249.
[9] Ratilainen 2007; 2010. In the English abstracts of the latest article, the word fireplace is misleadingly applied as a general term for all kinds of structures for keeping fire, also the word stove was applied.
[10] Campbell 2003, 13, 26–30; Filemio 2009, 579–586.
[11] Kroongaard Kristensen 2007, 230–232.
[12] Bonnier 1987, 26–27, 63–64; Bonnier et al. 2008, 311–314; Granberg 2006, 31–32; Konsmar 2013, 262; Redelius 2006, 101–104, 117–120.
[13] About Estonia: Bernotas 2013, 148–149, about Latvia: Sparitis 2007, 96–105.

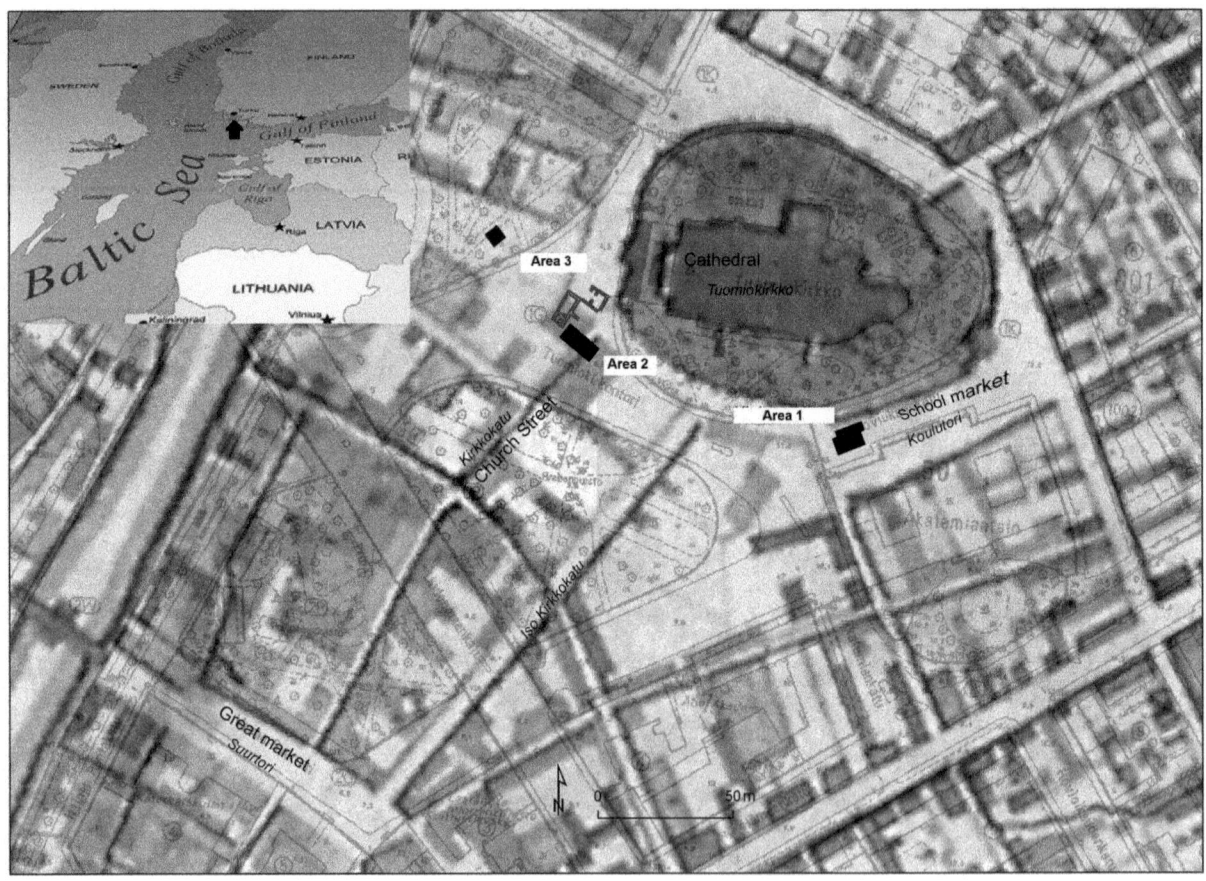

Figure 1. Turku, located in the Baltic Sea area (in the left corner, arrow), and the excavation areas 1–3 (solid black) in the current town plan over which the so-called Russian map (1743) was rectified. Church Street was one of the most important streets of the town in the Middle Ages leading from the Great Market to the Cathedral. (Large map made by Tapani Tuovinen and modified by Tanja Ratilainen. Original historical map in the archives of the National Board of Antiquities. Map of the Baltic area from: http://www.kamppissaitti.fi/a/user_images/LucrativoMarketingManagementOy/Case_3/isot/kartta.jpg)

end of the 13th century.[14] Recent studies have shown that brick was applied, more often both in ecclesiastical and in secular buildings, as well as in floors and heating systems, than thought before, although wood and corner jointed log building techniques prevailed in the Finnish medieval period.[15]

In Europe, the wattle and daub and cob techniques, which required raw clay, were applied in the Middle Ages.[16] Sometimes instead of cob, raw bricks were perhaps also used.[17] In timber-framed houses, clay, mixtures of clay and straw, or clay lumps struck in molds, were employed in the panels.[18] Clay lumps were also used in more modest house building.[19] However, none of these techniques included the use of lime mortar as a binder. The closest equivalent to the structure found in Turku was discovered in the excavations of the Skänninge Dominican convent in modern Sweden in 2005, where an edge of a brick foundation was built of unfired bricks.[20] In addition, unfired bricks relating to a medieval brick kiln have been recorded in Tartu, Estonia.[21]

The purpose of this paper is to present this peculiar foundation to a wider audience and to examine its function and purpose, since nothing exactly the same is so far known elsewhere in Europe. In addition, the structure has significant importance in the Finnish town archaeological research as it is the oldest brickwork structure in the area of the town dated with scientific dating methods.[22]

The structure and its stratigraphy

The earliest signs of human activity in Area 2 were the

---

[14] Drake 2007, 115. See also Harjula and Immonen 2012, 184.
[15] Drake 1991, 94–95; Hiekkanen 2001, 632; Ratilainen 2010; Seppänen 2012; Uotila 2002, 14–15; 2003, 131; 2009a, 82; 2009b, 45–46; Uotila et al. 2011, 299. Uotila. Traditional view: e.g. Ailio 1913, 1, 6; Lindberg 1919, 15–16; Gardberg 1957, 4–5, 20, 31; Valonen 1958, 21.
[16] Ayers, 2001, 36–37; Chapellot and Fossier 1985, 254–257; Schofield and Vince 2003, 128. Wattle and daub e.g.: Groothedde 2001, 181; Roesdahl and Scholkmann 2007, 160. Cob e.g.: Evans 2001, 53–59.
[17] Apparently without straw binder. Chapellot and Fossier 1985, 257. There is also evidence of using block-like clay layers in wells. (Pestell, e-mail to the author on 21st August 2013)
[18] Filemio 2009, 581–582; Schofield and Vince 2003, 128; Steuer 2007, 149; Spitzers 2001, 201–203.
[19] Ayers 2001, 39.
[20] Konsmar and Menander 2009; E-mail from Hanna Menander to the author December 21st, 2012 and December 23rd, 2012.
[21] Bernotas 2013, 143–145 and references therein.
[22] Ratilainen 2010.

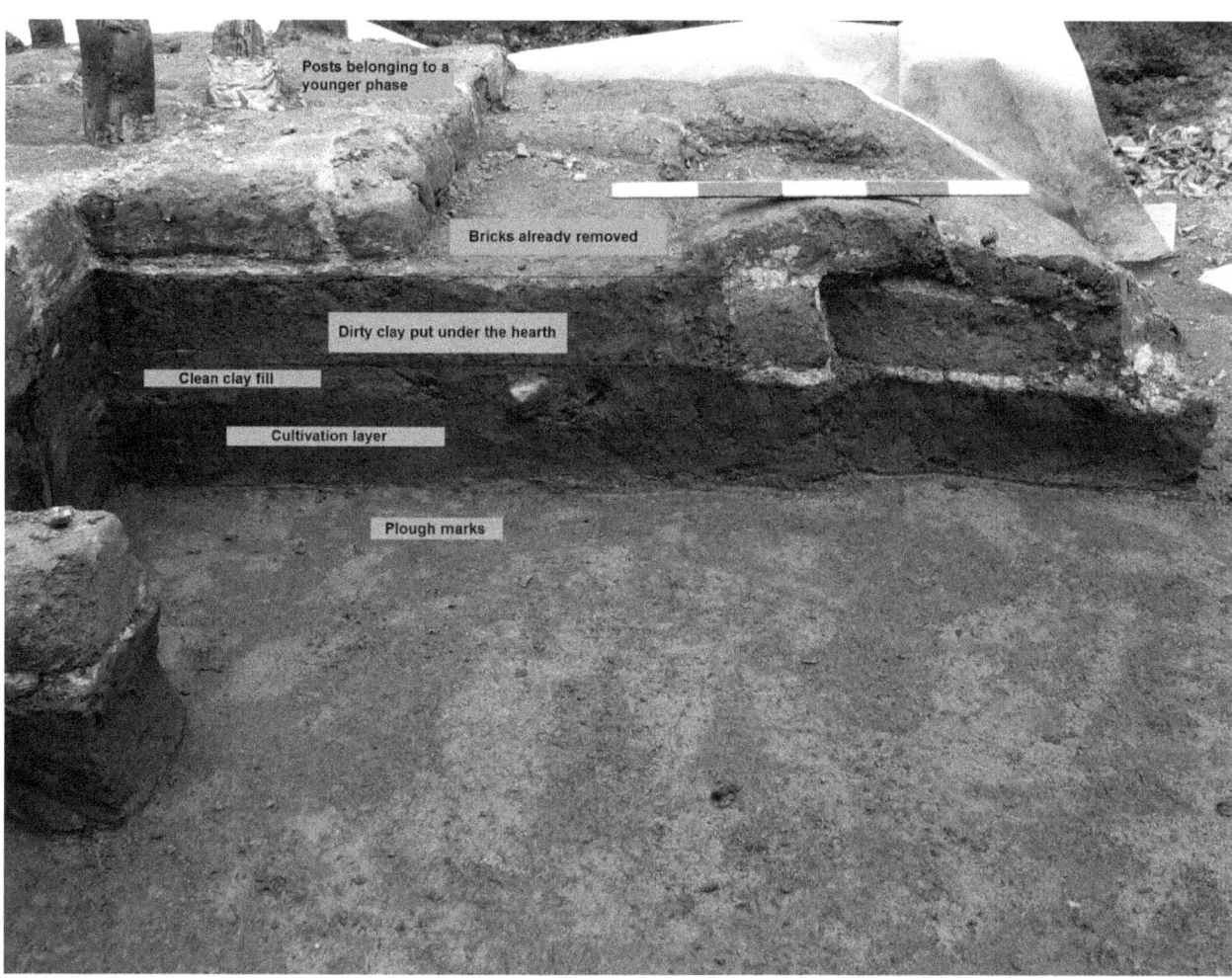

Figure 2. The NE-section of the structure showing the stratigraphy. Scale bar 50cm. Photo: Turku Museum Centre / Tanja Ratilainen.

plough marks in the untouched clay. Above them, there was a 10–15cm thick layer of clay colored by the organic matter, most probably the actual cultivation layer in which seeds of grain and of grain weeds were also discovered.[23] The structure made of unfired bricks was partly built directly on the cultivation layer and partly on the clean clay fill brought to the site (Figure 2). Under the structure, in the cultivation layer, several fragments of fully fired bricks were also recorded. A large round timber forming the SE-wall of the building was found partly against the SE-side of the structure made of unfired bricks. The oldest layer of Church Street was spread against that timber. Another timber forming the NW-wall was located on the opposite side. There were no signs of destruction by fire; therefore it is likely that the building was pulled down. In a later phase posts supporting the foundation of a stone house were struck through the structure in question[24] (Figure 3).

The size of the foundation was 2.2 x 2.2m. There were one to two courses of bricks left bound together with lime mortar. The size of the bricks varied between 26–33 x 12–18 x 6–10cm, the trimmean being 30 x 14 x 8cm. Cut pieces of unfired bricks or clay and lime mortar were laid on the edges (Figure 4). In the middle, there were bricks with slightly burned upper surface, likely the hearth, forming a round area c. 95cm in diameter.[25] Under the hearth, a layer of dirty clay and a hammer blade were found. On the edges of the hearth, bricks were cut lengthways to make it leveled. On the SW-side, half a width of a brick from the hearth, there was a rectangular area in which the upper surface of the raw bricks was colored black by charcoal. The size of the area was c. 75x25–35cm. On the SE-side of the foundation remains of badly conserved wood were discovered, which could hint that originally a timber box surrounded the structure. Outside the foundation there was a layer of burnt or heated sand. The layer covered an area of 70 x 70cm and it was 5–10cm thick. A shallow channel leading to the structure was also found. It measured 110 x 5–25 x 10–15cm and it was filled with clay and lime mortar[26] (Figure 5).

The lime mortar applied was still hard in some parts of

---

[23] Lempiäinen, 2010, 118–119; Ratilainen 2010, 32–33.
[24] Ratilainen 2010, 32–33.
[25] The area was c. three brick lengths long and six brick widths wide.
[26] Ratilainen 2010, 34.

Figure 3. Destruction layer spread over the foundation seen from the S. See the straight line of the SW edge of the layer showing the location of the timber at the time of the pulling down of the structure. NW- wall timber of the building, still at its original location; SE-timber's head is visible in the NE-profile. Photo: Turku Museum Centre / Elina Saloranta.

Figure 4. Detail (1/4) of the foundation seen from the SW. In the right corner the structure was likely pressed by a log surrounding it. In the upper left corner, the surface of the dirty clay layer located under the hearth. Photo: Turku Museum Centre / Elina Saloranta.

the foundation, but mostly what was left of it was loose sand colored by lime. The bricks were likely formed with a mould and dried out before they were used, otherwise they would have been impossible to handle. There was no straw or other organic material found in them; the bricks seemed to consist only of sand and clay. About 80 samples of bricks and of mortar were taken, but no further analyses have been carried out so far.[27]

The original size of the wooden building remained unclear as the NE-SW aligned timbers continued outside the excavation area. However, the size of the room was at least 17.5m2. Before dismantling the building, on the SW-side of the room a layer rich of organic material was formed. It included finds such as a wooden spoon, pieces of stave vessels, an entire small ceramic dish, unglazed greyware, stoneware, redware, nails, a tool and some leather waste. Also a ditch continuing into the NE-profile of the excavation area was discovered. The ditch was filled with organic debris. The demolition layer, containing mostly raw clay and pieces of burnt clay and lime mortar, was found above and in the surroundings of the foundation as well as in the ditch (Figure 3).

The foundation made of unfired bricks was most likely located in the S-corner of the room, since the SW-edge of the demolition layer was straight in a way that there must have been timber (perpendicular to the SE- and NW-timber) outlining the layer formed at the time of pulling down the structure [28] (Figure 3). The traces of corner-jointing found in the timbers support the interpretation (Figure 5).

Dating

Five radiocarbon samples of seeds found in the cultivation layer were analyzed. [29] The results showed that around the cathedral hill, cultivation activities were most probably practiced in the 13th century.[30] According to the dendrochronological dating results, the timbers set against and relating to the foundation were felled in 1288–1313 and 1317–1319.[31] There was no clear evidence for secondary use of the timbers. Based on the younger timber, the construction of the building began at the earliest at the end of 1310s. In addition, dendrochronological dating results from a stratigraphically younger building period showed that the first building went out of use likely by the mid-1340s, possibly already by 1342.[32] The pottery finds are also congruent with these dating results.[33] Therefore the building with its oven was most likely built in 1320s, at the earliest in spring 1317, and it was in use until the early 1340s.[34]

Discussion

It is often challenging to interpret the nature and the function of the structure based only on a discovered foundation. For example, open central hearths, which were constructed in Turku until the beginning of the 15th century, were usually built in the center of the room, without any foundations or use of lime mortar holding stones, surrounded by a timber box. These rectangular or round structures were relatively small. However, there are also cases in which structures interpreted as open hearths have been found in the corners or next to walls and there is evidence of constructing foundations as well.[35] The ovens, in which stones, bricks and lime mortar were applied, were built in the town since the 14th century. They were usually rectangular in plan and 2–4 m2 in size, with or without wooden foundations. They were usually larger than open hearths and located in the corner of the room, but there are exceptions to these as well.[36] As pointed out by Seppänen, it seems reasonable to assume that, besides the function of the oven, the size of the room to be heated affected its size.[37] When chimneys, open fireplaces, ovens with open fireplaces, or ovens with stoves emerged in the town is difficult to deduct based on the archaeological evidence, but most of them probably appeared during the late Middle Ages.[38]

Based on the size of the foundation, the presence of hearth, the likely location in the corner of the room, and the definite use of lime mortar, the structure has been interpreted as an oven. The area colored by charcoal and the burnt sand could suggest that the mouth of the oven was located on the NW-side of the structure and some kind of a stove may have been in front of it. The channel leading to the structure could indicate the need to produce extra heat with bellows during use. However, no slag, no pottery or no brick production waste was found in connection to it. The artefacts found in the building rather refer to ordinary household activities, though, perhaps of a well-off resident.[39] The location of the building by Church Street, near the Cathedral could support the idea that resident was member of the clergy. It is interesting that the ground in the room was not leveled, but inclined towards the NW-wall, which may be due to problems with water. [40]

The roundish shape of the hearth, the use of clay and the air channel could suggest that the structure was similar to the dome-shaped smoke-stoves or ovens found, for example, at the Raision Mulli site,[41] but no traces of the wattle and daub

---

[27] Ratilainen 2010, 34 and endnote 17.
[28] Ratilainen 2010, 33, 35.
[29] Report on C14 dating in the Poznan Radiocarbon Laboratory. Job nro: 2294/07, dated 14.8.2007.
[30] Pihlman 2010, 15–16. In English: Pihlman 2010, 22.
[31] Zetterberg 2006; 2007.
[32] Zetterberg 2007; Ratilainen 2010, 35.
[33] Pers. comm. Pihman 2007; Ainasoja et al. 2007.
[34] Ratilainen 2010, 35.

[35] Seppänen 2012, 709–720.
[36] No ovens with cold laid stone structures and loose top stones, in use until the 20th century in the countryside, have been found so far in medieval Turku, except for one structure perhaps relating to a sauna. Seppänen 2012, 709–720.
[37] Seppänen 2012, 720 and the references therein.
[38] Seppänen 2012, 723–727, 733–736.
[39] Ainasoja et al. 2007; Ratilainen 2007; 16; 2010, 34–35.
[40] There were no signs of a compact earthen/clay floor, but no signs of a wooden floor either (Ainasoja et al. 2007). Maybe there had been a wooden one, which was also removed and transferred to some other place for other function.
[41] e.g. Vuorinen 2003, 192–193; 2009, 85–86.

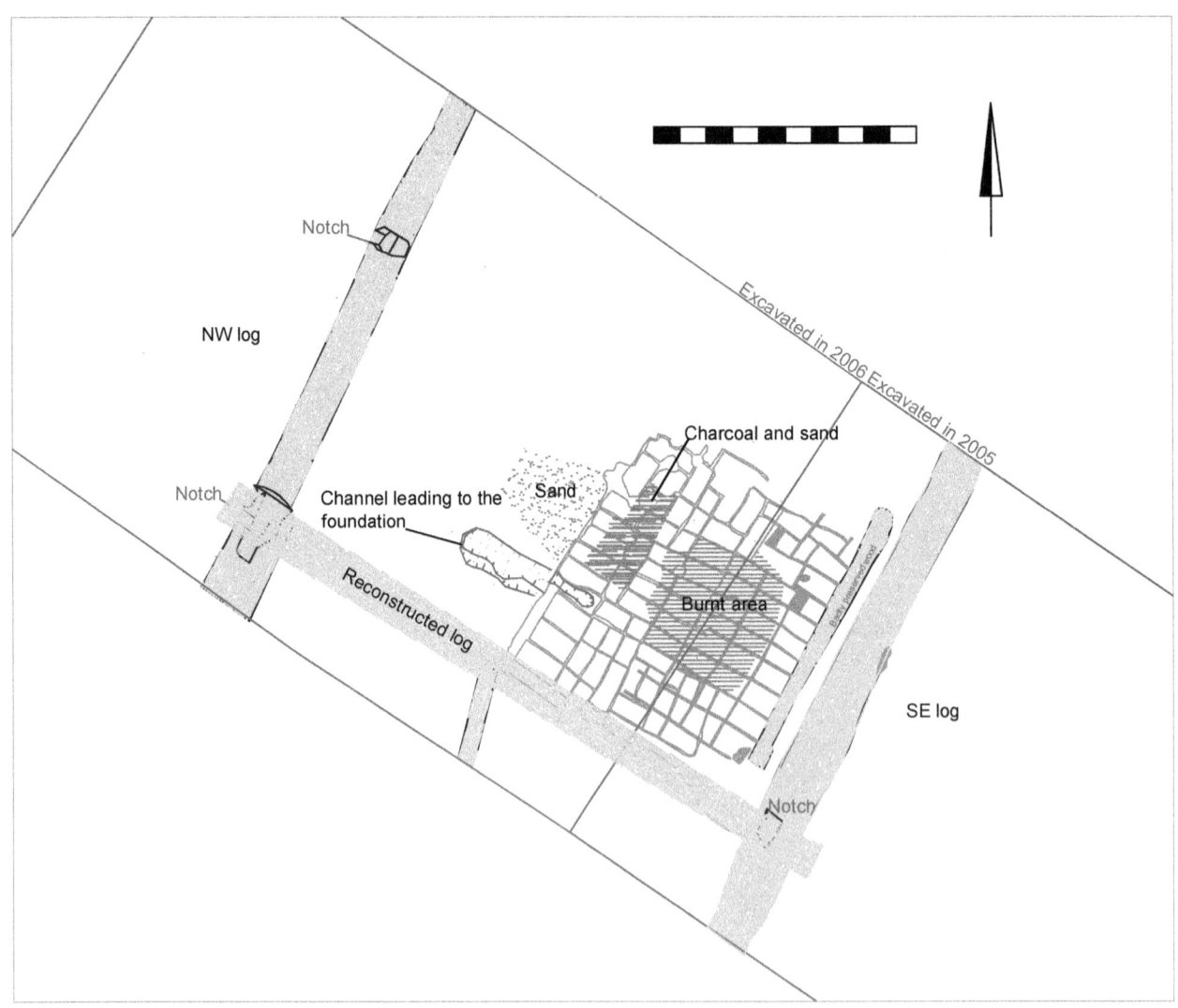

Figure 5. The building with its oven. Scale bar 2 meters. Map by Turku Museum Centre/ Mika Ainasoja and Tanja Ratilainen.

technique were discovered.[42] Therefore, I would consider it likely that the upper part of the structure was made of unfired bricks and lime mortar as well. The chamber around the hearth could have been shaped by cutting bricks and plastering them with clay; it might have even been vaulted.

The fact that the building with its oven was likely pulled down might suggest that, besides constructing with unfired bricks, the structure was built for temporary purposes. Moreover, there is no certainty that timber walls really ever surrounded it. In addition, the stratigraphic sequence of the units (firstly the oven, secondly the walls, thirdly the foundation of the street) could suggest that they all belong to different building phases. However, based on the fact that the structures were aligned with the street, in addition to the finds discovered in the layers relating to them, it is likely that they all were in use at the same time and compose the oldest phase of urban activities in the area. The evidence suggests that the builders knew the location of the street when they started to construct the oven and the walls around it.[43] The building with its oven was in use for about 20 years, which is a fairly ordinary time span in a mostly wooden town.[44] Therefore, it seems likely that the oven was not built for temporary purposes.

Finally, the foundation of an oven or a stove discovered in the Skänninge Dominican convent excavations and the brick kiln found in Tartu, Estonia have to be examined here.[45] The structure found in the kitchen of the convent measured 2.10 x 1.9m and it was slightly trapezoidal in plan. The eastern edge of the foundation was built of unfired bricks, one row laid length-ways, one row sideways, without any lime mortar. The rest of the structure was made of ordinary bricks, heavily affected by fire. The size of the unfired bricks was about 28 x 14cm and of the fired ones

---

[42] E.g. neither imprints of branches nor wooden strips in the bits of burnt clay nor traces of straw or dung.

[43] Ainasoja et al. 2007; Ratilainen 2007; 16; 2010, 32–35.
[44] Seppänen 2003, 96; 2012, 623 and note 9.
[45] For the latest excavation results of the convent see Hedvall et al. 2013. About the kiln: Bernotas 2013, 143–145.

24x12cm.[46] In the western part of the structure there was a shallow oval pit, which was filled with mortar. It was probably formed by some large artefact located there. The strongest signs of fire were around the pit. The structure was widely dated to 1275–1350.[47] In the brick kiln of Kitsas St. 1, Tartu, fired and unfired bricks were applied in the walls of the kiln measuring 4x6.2m. Based on other similar structures relating to brick production found in the suburb area of Tartu, it was dated to the end of the 13th or to the beginning of the 14th century.[48]

The size and the plan of the ovens of Turku and Skänninge are about the same, but actually no other common features exist between them. It could be interpreted that raw bricks were applied in both because bricks in general were needed, not because the type of oven or stove required it, just like in the brick kiln found in Tartu. Perhaps some dried bricks were left over from a large brick production and were decided to be used in these ovens or stoves. At least for the Skänninge convent, bricks must have been produced in large quantities at the time.[49]

For Turku, the situation could be the same, since fragments of bricks were discovered under the oven and in the cultivation layer as far as Area 1 on the southeast side of the cathedral. In addition, there are at least five other cases from other parts of the medieval town in which pieces of bricks have been recorded in layers likely dating to the first half of the 14th century.[50] Furthermore, besides the structure in question, bricks, not alone, but as part of the structure, were applied in at least two open hearths and one possible oven or stove, as well as in two floor structures dating to the same period.[51] It seems improbable that builders would have taken the trouble to make or to import just a few bricks for these structures. Instead, the evidence suggests that bricks were already produced in the town or nearby in large quantities in the beginning of 14th century.[52]

Conclusions

It is likely that the curious foundation in question – an oven – made of unfired bricks found in Turku, Finland was probably used for ordinary cooking and heating in a large timber house in the first building phase of the plot in the 1320s until 1340s. The location by Church Street, near the Cathedral, and the finds discovered in connection with the house show that the resident was likely well-off, perhaps, a member of clergy.

It seems probable that the raw bricks were left over or were the by-products of a large building project going on in the town in the beginning of 14th century and that they were simply used in the oven that needed to be built.

Acknowledgements

I would like to thank the Finnish Doctoral Program in Archaeology that made it financially possible to organize the session in the EAA meeting in Helsinki in 2012. I would like to express my gratitude to the Turku Centre for Medieval and Early Modern Studies for granting funds for the layout expenses of this publication. I thank Jean Price for English proofreading.

References

Ailio, J. 1913. Hattulan kirkko Hämeen emäkirkko. Kaikuja Hämeestä VIII:2–10.

Ainasoja, M., Harjula, J., Majantie, K., Pihlman, A., Ratilainen, T., Saloranta, E., Seppänen, L. and Tuovinen, T. 2007. Turku I, Tuomiokirkontori (Varhainen Turku -hanke). Kaupunkiarkeologiset tutkimukset 2005–2006. Unpublished excavation report.

Ayers, B. S. 2001. Domestic Architecture in Norwich from the 12th to the 17th Century. Lübecker Kolloquium zur Stadtarchäologie im Hanseraum III: Der Hausbau. Dahmen, B., Gläser, M., Oltmanns, U., Schindel, S. (eds), 35–48, Lübeck, Schmidt-Römhild.

Bernotas, R. 2013. Brick-making in the medieval Livonia – the Estonian example. Estonian Journal of Archaeology 17: 139–156. doi: 10.3176/arch.2013.2.03

Bonnier, A. C.1987. Kyrkorna berättar: Upplands kyrkor 1250–1350. Upplands Fornminnesförenings tidskrift 51. Upplands fornminnesförening, Uppsala.

Bonnier, A. C., G. Hägg ja I. Sjöström 2008. Svenska kyrkor: en historisk reseguide. Medström, Stockholm.

Campbell, J. W. P. 2003. Brick. A World History. Thames & Hudson, London.

Chapelot, J. and Fossier, R.1985. The Village & House in the Middle Ages. B. T. Batsford Ltd London, London.

Drake, K. 1991. Raseborg - grostensmurarna berättär sin historia. In: Snappertuna:en kustbygds hävder. Del 1, Forntid-1809, Rask, H. (ed.), 87–140, Ekenäs, Snappertuna sockenhistorie kommitté.

Drake, K. 2007. Gotische Backsteinbaukunst in Finnland. Backstein Baukunst. Zur Denkmalkultur des Ostseeraums. Dokumentation der Tagung zum 75. Geburtstag von Gottfiried Kiesow in der Wismarer St. Georgen-Kirche, 31.8–1.9.2006, 106–115, Wismar, Deutsche Stiftung Denkmalschutz.

Evans, D. H. 2001. Urban Domestic Architecture in the Lower Hull Valley in the Medieval and Early Post-Medieval periods. In: Lübecker Kolloquium zur Stadtarchäologie im Hanseraum III: Der Hausbau, Dahmen, B., Gläser, M., Oltmanns, U., Schindel, S. (eds), 49–76, Lübeck, Schmidt-Römhild.

Filemio, V. 2009. The Architecture and the Mechanics of Earthen Structures. Proceedings of the Third

---
[46] Length x width. No thickness was given. E-mail sent by Konsmar to the author November 15th, 2013.
[47] Konsmar and Menander 2009a, 23; Menander and Arcini 2013, 201–202; E-mail sent by Konsmar to the author November 15th, 2013; E-mails sent by Menander to the author December 20th and December 23rd, 2012. I find it interesting that even if most of the structure seems to have suffered from a heavy fire, the part in which there are unfired bricks did not.
[48] Bernotas 2013, 143–149.
[49] Menander and Arcini 2013, 201–210; Konsmar 2013, 261–263.
[50] Ratilainen 2010, 39–40.
[51] Ratilainen 2010, 37–38.
[52] Ratilainen 2010, 45–46.

International Congress on Construction History, May 2009, volume 2: 579–586.

Gardberg, C. J. 1957. Med murslev och timmerbila: drag ur det finländska byggnadshantverketshistoria. Helsingfors.

Gardberg, C. J. 1969. Turun keskiaikainen asemakaava. Turun kaupungin historiallinen museo,vuosijulkaisu 32–33: 5–52.

Gardberg, C. J. 1971. Turun kaupungin historia 1100-luvun puolivälistä vuoteen 1366. In: Turun kaupungin historia kivikaudesta vuoteen 1366; 117–315, Turku.

Graham-Campbell, J. 2007. Introduction. Archaeology of medieval Europe Volume 1, Eight to twelfth centuries AD, Graham-Campbell, J. and Valor, M. (eds), 11–18, Aarhus, Aarhus University Press.

Granberg, G. 2006. Uppsala stift: historiska perspektiv: om kyrka och tro i Uppland, Gästrikland och Hälsingland under tusen år. Stockholm, Verbum.

Groothedde, Michel. 2001. Der Hausbau in Zutphen zwischen 850–1400. Lübecker Kolloquium zur Stadtarchäologie im Hanseraum III: Der Hausbau, Dahmen, B., Gläser, M., Oltmanns, U., Schindel, S. (eds), 175–196, Lübeck, Verlag Schmidt-Römhild.

Harjula, J. and Immonen, V. 2012. Reviving old archaeological material. A project analysing finds from the early medieval fortified site of Koroinen in Turku. Castella Maris Baltici X: 183–190. Uotila, K., Mikkola, T. and Vilkuna, A.-M. (eds), Suomen keskiajan arkeologian seura, Saarijärvi.

Hedvall, R., Lindeblad, K., and Menander, H. (eds) 2013. Borgare, bröder och bönder: arkeologiska perspektiv på Skänninges äldre historia. Stockholm, Arkeologiska uppdragsverksamheten (UV), Riksantikvarieämbetet.

Helle, K. 2003. Introduction. The Cambridge history of Scandinavia. Vol. 1, Prehistory to 1520, Helle, K., (ed.), 1–12, Cambridge, Cambridge University Press.

Hiekkanen, M. 1997. Present and Future Urban Archaeology in Turku (Åbo). In Lübecker Kolloquium zur Stadtarchäologie im Hanseraum I: Stand, Aufgaben und Perspektiven, 377–385, Lübeck, Verlag Schmidt-Römhild.

Hiekkanen, Markus. 2001. Domestic Building Remains in Turku, Finland. In: Der Hausbau.Lübecker Kolloquium zur Stadarchäologie im Hanseraum III. Herausgegeben von Manfred Gläser, 627–633, Lübeck, Verlag Schmidt-Römhild.

Hiekkanen, M. 2002. The Christianization of Finland: a case of want of power in a peripheral area. In: Centre, region, periphery Volume 1: 488–497, Hertingen, Folio-Verlag Dr. G. Wesselkamp.

Konsmar, A. 2013. Tidiga tegelbyggare i Östergötland. In: Borgare, bröder och bönder – Arkeologiska perspektiv på Skänninges äldre historia. Rikard Hedvall, Karin Lindeblad and Hanna Menander (eds), 259–270, Stockholm, Arkeologiska uppdragsverksamheten (UV), Riksantikvarieämbetet.

Konsmar, A. 2013 E-mail to the author November 15th, 2013.

Konsmar, A. and Menander, H. 2009. S:t Olofs konvent. Skänningeprojektet I, RAÄ 20 Skänninge 2:1, 3:1 Skänninge stad, Mjölby kommun Östergötland Dnr 423–1717–2003. Excavation report available at: http://samla.raa.se/xmlui/handle/raa/6031.

Kroongaard Kristensen, H. 2007. The Production and Use of Bricks. In: Archaeology of medieval Europe Volume 1, Eight to twelfth centuries AD, Graham-Campbell, J. and Valor, M. (eds), 230–232, Aarhus, Aarhus University Press.

Lempiäinen, T. 2010. Hyöty- ja luonnonkasveja 1300-luvun Turussa – muutos maaseutuasutuksesta keskiaikaiseksi kaupungiksi. English summary: Useful and native plants in the 14th century Turku: From a rural settlement to a medieval town. In: Varhainen Turku. Raportteja 22: 95–123 , Söderström, M. (ed.), Turun museokeskus, Turku.

Lindberg, C. 1919. Om teglets användning i finska medeltida gråstenskyrkor. H. Schildt, Helsingfors.

Menander, H. 2012. Unfired bricks. E-mail to the author, December 20, 2012.

Menander, H. 2012. Unfired bricks. E-mail to the author, December 23, 2012.

Menander, H. and Arcini, C. 2013. Dominikankonventet S:t Olof. In: Borgare, bröder och bönder – Arkeologiska perspektiv på Skänninges äldre historia. Hedvall, R., Lindeblad, K. and Menander, H. (eds), 191–228, Stockholm, Arkeologiska uppdragsverksamheten (UV), Riksantikvarieämbetet.

Pestell, Tim. 2013. E-mail sent to the author August 21st, 2013.

Pihlman, Aki. 2007a. Varhainen Turku -tutkimushanke. Uusia arkeologisia tulkintoja Turun kaupungin muodostumisesta. HIT History in Turku: Tietoja, taitoja ja löytöjä. Näyttely Turun linnassa 15.6. – 23.9.2007. Turun maakuntamuseo näyttelyesite 42: 10–15.

Pihlman, Aki. 2007b. The Archaeology of Early Turku and the Extent of the town. The European Hansa. Report 21: 85–94. Turun maakuntamuseo, Turku.

Pihlman, Aki. 2007. pers. comm.

Pihlman, Aki. 2010. Turun kaupungin muodostuminen ja kaupunkiasutuksen laajeneminen 1300-luvulla. In: Varhainen Turku. Raportteja 22: 9–29, Söderström, M. (ed.). Turun museokeskus, Turku.

Pihlman, Aki and Kostet, Juhani. 1986. Turku. Keskiajan kaupungit 3. Turun maakuntamuseo,Turku.

Ratilainen, Tanja. 2007. Kirkkokadun ja sen varrella sijainneen tontin vaiheita. In: HIT – History in Turku: tietoja, taitoja ja löytöjä. Näyttely Turun linnassa 15.6.–23.9.2007. Turun maakuntamuseo, näyttelyesite 42: 16–23.

Ratilainen, T. 2010. Tiilen käytöstä 1300-luvun Turussa. English summary: Brick use in Turku in the 14th century. In: Varhainen Turku. Raportteja 22: 31–55, Söderström, M. (ed.), Turun museokeskus, Turku.

REA 1890. Registrum Ecclesiæ Aboensis, eller, Åbo domkyrkas Svartbok: med tillägg ur Skoklosters Codex Aboensis. Edited by Hausen, Reinhold, J. Simelii arfvingars boktryckeri, Helsingfors. Also available online: http://extranet.narc.fi/DF/detail.php?id=72.

Report on C-14 dating in the Poznan Radiocarbon Laboratory. Job nro: 2294/07, dated August 14th, 2007. Appendix to the unpublished excavation report by Ainasoja et al. 2007, Archives of the Turku Museum Centre.

Roesdahl, E. and Scholkmann, B. 2007. Housing culture. In: Archaeology of medieval Europe Volume 1, Eight to twelfth centuries AD, Graham-Campbell, J. and Valor, M. (eds), 154–180, Aarhus, Aarhus University Press.

Ruuth, J. W. 1909. Åbo stads historia under medeltiden och 1500-talet. Första häftet. Bidrag till Åbo stads historia. Utgifna på föranstaltande av bestryrelsen för Åbo stads historiska museum, andra serien, IX, Helsingfors.

Redelius, G. 2006. Sigtuna och Sko. En tid, ett rum. Gunnar Redelius, Uppsala.

Saloranta, E. 2010. Puurakentaminen ja puurakennukset Turussa 1300-luvulla. In: Varhainen Turku. Raportteja 22: 57–77, Söderström, M. (ed.) Turun museokeskus, Turku,

Schofield, J. and Vince, A. 2003. The medieval towns. The Archaeology of British Towns in the European Setting. Continuum, London, New York.

Seppänen, Liisa. 2003. Salvoksista sosiaalisiin systeemeihin. Kaupunkia pintaa syvemmältä. Arkeologisia näkökulmia Turun historiaan. Seppänen, L. (ed.) Archaeologia Medii Aevi Finlandiae IX: 89–104, Suomen keskiajan arkeologian seura, Turku.

Seppänen, Liisa. 2012. Rakentaminen ja kaupunkikuvan muutokset keskiajan Turussa. Erityistarkastelussa Åbo Akademin päärakennuksen tontin arkeologinen aineisto. Dissertation: Turun yliopisto. Turku, Turun yliopisto. http://www.doria.fi/handle/10024/86116.

Siiriäinen, A. 1987. Suomen esihistoria. In Suomen historian pikkujättiläinen, Zetterberg, S. (ed.), 11–39, WSOY, Porvoo.

Sparitis, O. 2007. Backsteingotik im Baltikum. In: Backstein Baukunst. Zur Denkmalkultur des Ostseeraum, 96–105, Deutsche Stiftung Denkmalschutz, Wismar.

Spitzers, Thomas A. 2001. Archaeological Data on Domestic Architecture in Deventer. Lübecker Kolloquium zur Stadtarchäologie im Hanseraum III: Der Hausbau, Dahmen, B., Gläser, M., Oltmanns, U., Schindel (eds), 197–212, Lübeck, Verlag Schmidt-Römhild.

Steuer, Heiko. 2007. Part 2: Central, Northern, Eastern and Southern Europe. In: Archaeology of medieval Europe Volume 1, Eight to twelfth centuries AD, Graham-Campbell, J. and Valor, M. (eds), 129–153, Aarhus, Aarhus University Press.

Uotila, Kari 2002. Åbo medeltida rådhus – en stenbyggnad vid ett medeltida torg. META: medeltidsarkeologisk tidskrift 4/2002: 3–18.

Uotila, K. 2003. Kivitaloja keskiajan Turussa. In: Kaupunkia pintaa syvemmältä. Arkeologisia näkökulmia Turun historiaan. Archaeologia Medii Aevi Finlandiae IX, Seppänen, L. (ed.), 121–134, Suomen keskiajan arkeologian seura, Turku.

Uotila, K. 2009a. Keskiaikainen Hämeen linna arkeologisena tutkimushaasteena. Arx Tavastica: Hämeenlinnan historiallisen seuran julkaisu (13):77–95.

Uotila, K. 2009b. Rauniokohde ja sen tutkimus. In: Ikuinen raunio, Muhonen, T. and Lehto-Vahtera, J. (eds), 42–63, Turku, Matti Koivurinnan säätiö.

Uotila, K. Helama, S., Holopainen, J., Hilasvuori, E. and Oinonen, M. 2011. Naantalin luostarin rakentaminen 1400-luvun kuluessa. Naantalin luostarin rannassa: arkipäivä Naantalin luostarissa ja sen liepeillä - Stranden vid Nådendals kloster: vardagen i klostret och dess omgivning. Uotila, K. (ed.), 297–302, Kaarina, Muuritutkimus.

Valonen, N. 1958. Turun viemärikaivantolöydöistä. Vuosijulkaisu / Turun kaupungin historiallinen museo (20–21): 12–110.

Törnblom, L. 1993. Medeltiden. In: Finlands historia 1: 271–426. Schildt, Esbo.

Virrankoski, P. 2009. Suomen historia 1. 2. korj. p. ed. Suomalaisen Kirjallisuuden Seura, Helsinki.

Vuorinen, Juha-Matti. 2003. Hallitalo ja hirsirakennus – elämää ahtaassa asumuksessa mutta väljässä pihapiirissä. Muinainen Kalanti ja sen naapurit. Talonpojan maailma rautakaudelta keskiajalle, Kaitainen, V., Laukkanen, E. and Uotila, U. (eds), 185–198, Helsinki, Suomalaisen kirjallisuuden seura.

Vuorinen, Juha-Matti. 2009. Rakennukset ja rakentajat Raision Ihalassa rautakauden lopulla javarhaisella keskiajalla. Turun yliopiston julkaisuja, Annales Universitatis Turkuensis, sarja C, osa 281, Scripta Lingua Fennica Edita. Turun yliopisto, Turku. https://www.doria.fi/handle/10024/45215

Zetterberg, Pentti. 2006. Varhaisinta Turkua -hankkeen arkeologisten kaivausten puulöytöjen iänmääritys, dendrokronologiset ajoitukset FIT0001–FIT0014. Joensuun yliopisto, Biotieteiden tiedekunta. Ekologian tutkimusinstituutti, Dendrokronologian laboratorio, ajoitusseloste 280. Appendix to the unpublished excavation report by Ainasoja et al. 2007, Archives of the Turku Museum Centre.

Zetterberg, Pentti. 2007. Varhainen Turku -hankkeen arkeologisten kaivausten puulöytöjen iänmääritys, dendrokronologiset ajoitukset FIT0015–FIT0026. Joensuun yliopisto, Biotieteiden tiedekunta. Ekologian tutkimusinstituutti, Dendrokronologian laboratorio, ajoitusseloste 307. Appendix to the unpublished excavation report by Ainasoja et al. 2007, The archives of the Turku Museum Centre.

www.ingramcontent.com/pod-product-compliance
Lightning Source LLC
Chambersburg PA
CBHW041708290426
44108CB00027B/2891